Edexcel

Foundation

GCSE Modular Mathematics

4-Speed Revision Guide

Keith Pledger

Gareth Cole

Peter Jolly

Graham Newman

www.heinemann.co.uk
- Free online support
- Useful weblinks
- 24 hour online ordering

Heinemann is an imprint of Pearson Education Limited, a company incorporated in England and Wales, having its registered office at Edinburgh Gate, Harlow, Essex, CM20 2JE. Registered company number: 872828

www.heinemann.co.uk

Heinemann is a registered trademark of Pearson Education Limited

Text © Keith Pledger, Gareth Cole, Peter Jolly and Graham Newman, 2007

First published 2007

12 11 10 09 08 07
10 9 8 7 6 5 4 3 2 1

British Library Cataloguing in Publication Data is available from the British Library on request.

ISBN 978 0 435807 184

Typeset by Tech-Set Ltd, Gateshead, Tyne and Wear
Cover design by Tony Richardson
Cover photo/illustration © Digital Vision
Printed in the UK at Scotprint

Acknowledgements

Every effort has been made to contact copyright holders of material reproduced in this book. Any omissions will be rectified in subsequent printings if notice is given to the publishers.

Welcome to 4-speed revision!

4-speed revision lets you plan your revision at the speed you need.

- 1st speed pages are **green** – use these to revise topics thoroughly.

- 2nd speed pages are **orange** – use the topic tests on these pages to identify your weaknesses in detail.

- 3rd speed pages are **blue** – use the subject tests on these pages to help you identify the topics you need to work on the most.

- 4th speed pages are red – use these to check you know the key facts.

Pages iv to vi show you how the different speed pages work – so you can choose your speed.

Contents

Revising for your GCSE maths exam

WHY? Because you need to be prepared for every question in the exam.

WHEN?

Sooner rather than later.

Make yourself a Revision Timetable and stick to it.

Set revision sessions of manageable length before and after school, say 10–15 minutes each to start with.

Build up to longer sessions, with short breaks every 25 minutes or so.

Don't cram the night before.

WHERE?

Choose a place where
- you will not be disturbed
- you can really concentrate
- you have all the equipment you need

WHAT?

Check which exam you are revising for – Unit 1, Unit 2 Stage 1, Unit 2 Stage 2, or Unit 3.

Use the revision planner on pages viii–xii to work out which pages you need to revise.

1st speed – green pages – thorough revision

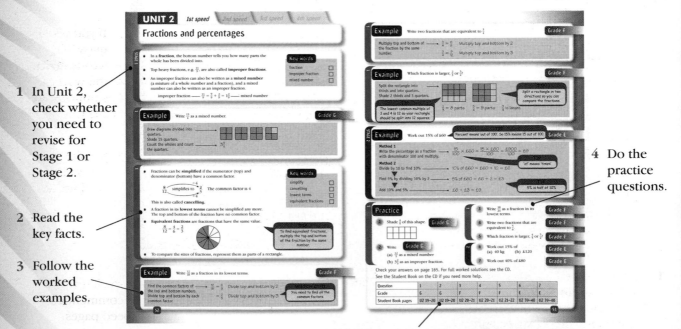

1 In Unit 2, check whether you need to revise for Stage 1 or Stage 2.

2 Read the key facts.

3 Follow the worked examples.

4 Do the practice questions.

5 For more help with tricky topics, read the recommended Student book pages on the CD.

2nd speed – orange pages – identify your weaknesses in detail

1 Take a topic test.

2 Check your answers against the worked solutions.

3 See the grade you are working at.

4 If you get any questions wrong, work through the recommended 1st speed pages.

UNIT 1 — 1st speed | 2nd speed | 3rd speed | 4th speed

Presenting data: topic test

Check how well you know this topic by answering these questions.
First cover the answers on the facing page.

Test questions

1. Jez has a stamp collection. The pictogram shows the numbers of stamps he has from France, Germany and Italy.
   (a) Write down the number of stamps from France.
   (b) Write down the number of stamps from Germany.
   Jez has 60 stamps from Spain and 30 stamps from Austria.
   (c) Use this information to complete the pictogram.

France	
Germany	
Italy	
Spain	
Austria	

   Key: represents 20 stamps

2. This accurate pie chart shows information about the medals won by the UK in the Paralympic Games in Athens 2004.
   The total number of medals won by the UK was 94.
   (a) Find the number of gold, silver and bronze medals.
   (b) What fraction of the total medals won were silver?

3. An internet company recorded the number of orders it received on each of 30 days. Here are the results:

18	48	55	12	43	26	40	14	26	16
26	15	58	39	56	15	58	57	16	58
29	44	29	26	44	26	51	52	24	15

   Represent this data using an ordered stem and leaf diagram. Include a key.

4. The table shows the test marks for eight students.

   | Maths | 25 | 6 | 17 | 33 | 21 | 10 | 17 | 28 |
   | Science | 20 | 8 | 15 | 29 | 22 | 9 | 19 | 30 |

   (a) Draw a scatter graph to show this information.
   (b) Describe the relationship between the two sets of data.
   (c) Draw a line of best fit on your scatter graph.
   (f) Fatima's maths mark was 15. Use your line of best fit to estimate her science mark.

5. Toby measures the weights of the tomatoes from his tomato plants. His results are summarised in the table.
   (a) On one set of axes, draw two frequency polygons to show this information.
   (b) Comment on Toby's plants.

Weight (g)	Frequency plant A	Frequency plant B
30–39	1	3
40–49	2	7
50–59	3	9
60–69	6	7
70–79	5	4
80–89	5	2

   Now check your answers – see the facing page.

Cover this page while you answer the test questions opposite.

Worked answers — Revise this on...

1. (a) 4 × 20 = 80 stamps — page 8
   (b) 2 × 20 + 10 = 50 stamps (c)

   | Spain | |
   | Austria | |

   Key: represents 20 stamps

2. 94 medals are represented by 360°. — page 10
   (a) Gold: × 94 = 35 medals (b)
   Silver: × 94 = 30 medals
   Bronze: × 94 = 29 medals

3. — page 11
   1 | 2, 3, 3, 4, 5, 6, 6, 8
   2 | 4, 6, 6, 6, 6, 6, 6, 9
   3 | 5, 6, 8, 9, 9
   4 | 0, 3, 4, 4, 8
   5 | 1, 2, 7, 8   Key: 2 | 4 means 24

4. (a), (c) — pages 12–13
   (b) The higher the maths mark, the higher the science mark. It is positive correlation.
   (c) See graph above. (d) 15   **Grade C**

5. (a) — pages 8–9
   (b) Most tomatoes weigh between 50 g and 69 g. Plant A has heavier tomatoes. Plant B has more tomatoes.

Tick the questions you got right.

Question	1	2	3	4ab	4cd	5
Grade	G	F/E	D	D	C	C

Mark the grade you are working at on your revision planner on page viii.

3rd speed – blue pages – focus on your top priority topics

1 Take a subject test – there is a test for Handling data in Unit 1, and one for each of Number, Algebra and Shape, space and measure in Units 2 and 3.

2 Check your answers.

3 If you get any questions wrong check the worked solutions on the CD.

4 See the grade you are working at.

5 Then work through the recommended 1st speed pages.

UNIT 1 — 1st speed | 2nd speed | 3rd speed | 4th speed

Handling data: subject test

Exam practice questions

1. Salih asked his friends 'What is your favourite sport?' Here are his results:

   soccer golf soccer rugby cricket
   rugby soccer soccer golf soccer
   soccer rugby cricket rugby soccer
   golf soccer rugby soccer rugby

Sport	Tally	Frequency
Soccer		
Golf		
Rugby		
Cricket		

   (a) Complete the table to summarise Salih's results.
   (b) Write down the number of friends whose favourite sport was rugby.
   (c) What was the most popular sport among his friends?

2. The pictogram shows some information about the numbers of DVDs rented from a petrol station.
   (a) Write down the number of DVDs rented on
   (i) Saturday (ii) Sunday.
   (b) 40 DVDs were rented on Monday, and 50 DVDs on Tuesday. Show this information on the pictogram.

Saturday	
Sunday	
Monday	
Tuesday	

   Key: represents 20 DVDs

3. Here is a list of Carol's test marks.
   7, 5, 8, 8, 9, 6, 8, 8, 6, 5
   (a) Write down the mode.
   (b) Work out the mean.
   (c) Work out the range.

4. This table gives information about 90 people's eye colour.

Eye colour	Number of people
Blue	40
Grey	15
Green	25
Brown	10

   Draw an accurate pie chart to show this information.

5. 20 people do a lap round a race track. Here are their times to the nearest second.
   62 46 39 55 28 44 65 41 48 57
   36 49 51 46 39 27 60 50 45 53
   (a) Draw an ordered stem and leaf diagram to show this information. Include a key.
   (b) Use your stem and leaf diagram to write down the median.

6. 100 students were asked how they came to school that day. Some of the results are shown in the two-way table.

	Car	Walk	Cycle	Total
Year 7		15	9	41
Year 8	5			22
Year 9		18		
Total	36		21	100

   (a) Complete the two-way table.
   (b) One of these students is picked at random. Write down the probability that this student
   (i) came to school by car (ii) is in Year 9 and cycled to school.

7. A freezer contains four flavours of ice-cream – vanilla, chocolate, strawberry and mint. The table shows the probabilities that Marcus chooses vanilla, chocolate or mint.

Flavour	Vanilla	Chocolate	Strawberry	Mint
Probability	0.2	0.4		0.15

   Work out the probability that he picks a strawberry ice-cream.

8. The table shows the heights and weights of ten students.

   | Weight (kg) | 75 | 65 | 82 | 76 | 71 | 65 | 77 | 70 | 72 | 68 |
   | Height (cm) | 185 | 182 | 191 | 188 | 184 | 166 | 175 | 178 | 181 | 180 |

   (a) Draw a scatter graph to show this information.
   (b) What type of correlation do you find?
   (c) Draw a line of best fit.
   (d) Use your scatter graph to estimate
   (i) the weight of a student whose height is 188 cm
   (ii) the height of a student whose weight is 74 kg.

   Check your answers on page 163–164. For full worked solutions see the CD.
   Tick the questions you got right.

Question	1	2	3a	3bc	4	5	6a	6b	7	8abc	8d
Grade	G	G	G	F	E	D	E	D	D	D	C
Revise this on page	2	8	16	10	11	23	23			12	

   Mark the grade you are working at on your revision planner on page viii.
   Go to the pages shown to revise for the ones you got wrong.

4th speed – red pages – instant overview of the key facts

1 Check that you understand the key facts for each subject.

2 Read the 1st speed pages of any topics you are not sure about.

3 Learn any formulae that are not on the formulae sheet.

UNIT 1 · 1st speed · 2nd speed · 3rd speed · **4th speed**

Handling data

Collecting and organising data

- A **tally chart** is a way of recording and displaying data.

Flavour	Tally	Frequency
Chocolate	⦀⦀ ⦀⦀⦀	9
Fruit	⦀⦀ ⦀	6
Lemon	⦀⦀	2
Banana	⦀⦀⦀	3

- **Two-way tables** are used to record or display information that is grouped in two categories.

Presenting data

- A **pictogram** uses symbols or pictures to represent quantities. It needs a **key** to show what one symbol represents.

Monday	
Tuesday	
Wednesday	
Thursday	
Friday	

Key: ▱ represents 10 bags of toffees

- A **bar chart** shows data that can be counted. You must leave a gap between the bars.
- A **dual bar chart** compares two sets of data.
- A **pie chart** is a way of displaying data when you want to show how something is shared or divided. The angles at the centre of a pie chart add up to 360°.

- A **stem and leaf diagram** shows the shape of a distribution and keeps all the data values. It needs a **key** to show how the stem and leaf are combined.

```
0 | 9
1 | 2, 3
2 | 0, 7
3 | 4, 8
4 | 1
```
Key: 1 | 2 means 12

- A **line graph** can be used to show continuous data.

Averages and the range

- The **mode** of a set of data is the value which occurs most often.
- The **median** is the middle value when the data are arranged in order of size.
- The **mean** of a set of data is the sum of the values divided by the number of values.
- The **range** of a set of data is the difference between the highest value and the lowest value.
- With a **frequency table**:

  $$\text{mean} = \frac{\sum fx}{\sum f}$$
  the sum of all the $(f \times x)$ values in the table / the sum of the frequencies

- For **grouped data**:
  – the **modal class** is the class interval with the highest frequency
  – you can state the **class interval** that contains the median
  – you can calculate an estimate of the **mean** using the middle value of each class interval.

Probability

- Probabilities can be shown on a **probability scale**.

  0 ⟶ ½ ⟶ 1
  impossible · unlikely · even chance · likely · certain

- If all the outcomes are equally likely,

  $$\text{probability} = \frac{\text{number of successful outcomes}}{\text{total number of possible outcomes}}$$

- $$\text{Estimated probability} = \frac{\text{number of successful trials}}{\text{total number of trials}}$$

Practice examination papers

1 Do the practice examination papers for the unit you need.

2 Use the 'Maths language in exams' page to decode what the question is asking you for.

3 For each question, how many marks is it worth? Make sure that you show all your working out to get all the marks.

4 Check your answers.

5 Check the worked solutions on the CD for any you got wrong and work through the recommended 1st speed pages.

Unit 2 Examination practice paper
A formula sheet can be found on page 161.
Stage 1 (multiple choice)
Calculator

1 A golf tournament had 5678 spectators. Which of the numbers below shows the number 5678 to the nearest ten?
A 5700 B 5678 C 5670
D 5800 E 5680

2 Here is part of a train timetable.

Manchester	06 30	09 00	09 40
Levenshulme	08 38	09 08	09 48
Heaton Chapel	06 42	09 12	09 52
Stockport	06 47	09 17	09 57

When will the 09 40 train from Manchester arrive at Heaton Chapel?
A 09 48 B 09 12 C 09 17
D 09 52 E 09 57

3 61 47 72 53 32
If these numbers are put in order from lowest to highest number, which would be the middle number?
A 61 B 47 C 72 D 53 E 32

4 What type of angle is this?
A Acute
B Obtuse
C Reflex
D Right-angled E Opposite

5 What is the value of the 4 in the number 24 752?
A 4 B 40 C 400
D 4000 E 40 000

6 What is the reading on the number line?

A 3.4 B 3.8 C 3.45 D 3.9 E 3.95

7 This is a series of patterns made from dots.

Pattern 1   Pattern 2   Pattern 3

| Pattern number | 1 | 2 | 3 |
| Number of dots | 6 | 9 | 12 |

How many dots are there in Pattern 4?
A 14 B 15 C 16 D 17 E 18

8 What is the size of the angle marked x.
Diagram NOT accurately drawn
A 10° B 20° C 30° D 40° E 50°

9 2°C −5°C 3°C −8°C 6°C
If these temperatures are written in order with the lowest temperature first, which temperature would be 4th in order?
A 2°C B −5°C C 3°C D −8°C E 6°C

10 Here are the first five terms in a sequence of numbers.
7 11 15 19
Which is the 9th term
A 27 B 35 C

11

What are the coordinates
A (4, −2) B (2,
D (4, 2) E (−

Unit 3 Examination practice paper
A formula sheet can be found on page 161.
Section A (calculator)

1 The cost of having a car serviced is £56.40 before VAT at 17½% is added.
Find the total cost after VAT is added. **(3 marks)**

2 Work out the value of y.

28 cm

53 cm

y cm

(3 marks)

3 The diameter of a wheel is 70 centimetres. Work out how many revolutions the wheel makes when travelling 1 kilometre. **(4 marks)**

4 Here is a list of ingredients for making Lemon Surprise for 4 people:
2 lemons   2 eggs
50 g butter   250 ml milk
100 g sugar   50 g self raising flour
Work out how much of each ingredient is needed to make Lemon Surprise for 10 people. **(3 marks)**

5 (a) Find the value of n in $\frac{a^6 \times a^4}{a^2} = a^n$
(b) Simplify $(2x^3y)^3$ **(4 marks)**

6 Reflect the shape P using the dotted line as the mirror line.

P

(2 marks)

7 $s = \frac{1}{2}(u + v)t$ is a formula linking distance, speed and time.
(a) Work out the value of s when $u = 0$, $v = 6$ and $t = 3$
(b) Make v the subject of the formula. **(5 marks)**

8 Work out the area of a circle with radius 1.7 metres.
Give your answer correct to 5 significant figures. **(2 marks)**

9 $x^2 + 3x − 16 = 0$
Use trial and improvement to find the positive solution of this equation.
Give your answer correct to 1 decimal place. **(4 marks)**

10 Convert the fraction $\frac{5}{9}$ to a decimal.
Give your answer correct to 2 decimal places. **(2 marks)**

11 The total surface area, including the base, of this triangular prism is 2100 cm².

29 cm

20 cm

x cm

21 cm

Work out the value of x. **(4 marks)**

12 Work out the value of $5y^2 − 2y$ when $y = −4$ **(2 marks)**

13 Construct a triangle ABC with AB = 8 cm, angle A = 75° and angle B = 30° **(2 marks)**

⊙ CD

- Full text book for each unit. Use them to revise topics in more depth.

- Worked solutions to practice questions, practice exam papers and 3rd speed subject test questions.

Revision planner (pages viii–xii)

The revision planner shows the contents of each page.
You can use it to help you plan your revision:

1 Tick the topics you understand.

UNIT 2 (cont.)

Speed	Topic		I need to revise	I understand ✓
1st	**Coordinates**	page 62		
	2-D coordinates			
	3-D coordinates			
1st	**Algebraic line graphs**	page 64		
	Drawing straight-line graphs			
	Using straight-line graphs			
2nd	**ALGEBRA, SEQUENCES AND COORDINATES topic test**		I am working at grade ◯	
3rd	**UNIT 2 ALGEBRA subject test**		I am working at grade ◯	
1st	**Naming and calculating angles**	page 70		
	Naming angles			

2 After each test, mark on the grade you are working at.

Maths language in exams (page xiii)

Use this page to decode what a question really means.

Maths language in exams

When a question says...	What it means
You must show your working...	You will lose marks if you do not show how you worked out the answer.
Estimate...	Usually means round numbers to 1 significant figure and then carry out the calculation.
Calculate...	Some working out is needed – so show it!
Work out OR Find...	A written or mental calculation is needed.
Write down...	Written working out is not usually required.
Give an exact value of...	No rounding or approximations.
Give your answer to an appropriate degree of accuracy...	If the numbers in the question are given to 2 decimal places, give your answer to 2 decimal places.
Give your answer in its simplest	Usually means you will need to cancel a fraction or a ratio.

Revision planner

Use this to help you plan your revision.

- Tick the topics you understand ✓
- After each test, write in the grade you are working at Ⓒ

UNIT 1

Speed	Topic		I need to revise	I understand ✓
1st	**Collecting data**	page 2		
	Tally charts			
	Questionnaires			
1st	**Organising data**	page 4		
	Tables			
	Two-way tables			
2nd	**COLLECTING AND ORGANISING DATA topic test**		I am working at grade ◯	
1st	**Charts**	page 8		
	Pictograms and bar charts			
	Histograms and frequency polygons			
1st	**Pie charts and stem and leaf diagrams**	page 10		
	Pie charts			
	Stem and leaf diagrams			
1st	**Time series and scatter graphs**	page 12		
	Line graphs and scatter graphs			
2nd	**PRESENTING DATA topic test**		I am working at grade ◯	
1st	**Averages and the range**	page 16		
	Mode			
	Median			
	Mean			
	Range			
2nd	**AVERAGES AND THE RANGE topic test**		I am working at grade ◯	
1st	**Probability**	page 22		
	Probability scale			
	Single events			
	Combined events			
	Estimated probability/relative frequency			
2nd	**PROBABILITY topic test**		I am working at grade ◯	
3rd	**Handling Data subject test**		I am working at grade ◯	
4th	**Unit 1 key facts**			
	Unit 1 Examination practice paper			

UNIT 2

UNIT 2 (cont.)

Speed	Topic	I need to revise	I understand ✓
1st	**Coordinates** page 62		
	2-D coordinates		
	3-D coordinates		
1st	**Algebraic line graphs** page 64		
	Drawing straight-line graphs		
	Using straight-line graphs		
2nd	**ALGEBRA, SEQUENCES AND LINE GRAPHS topic test**	I am working at grade ◯	
3rd	**UNIT 2 ALGEBRA subject test**	I am working at grade ◯	
1st	**Naming and calculating angles** page 70		
	Naming angles		
	Angles on a straight line, at a point, vertically opposite		
1st	**Working with angles** page 72		
	Corresponding and alternate angles		
2nd	**ANGLES topic test**	I am working at grade ◯	
1st	**Units of measurement** page 76		
	Metric units/conversion		
	Metric/imperial conversion		
	Speed		
	Timetables and 24-hour times		
2nd	**MEASURE topic test**	I am working at grade ◯	
1st	**Perimeter and area** page 80		
	Perimeter		
	Area		
	Surface area		
1st	**Volume, capacity and density** page 82		
	Volume, capacity and density		
	Prisms and cylinders		
2nd	**PERIMETER, AREA AND VOLUME topic test**	I am working at grade ◯	
3rd	**UNIT 2 SHAPE, SPACE AND MEASURE subject test**	I am working at grade ◯	
4th	**Unit 2 key facts**		
	Unit 2 stage 1 Examination practice paper		
	Unit 2 stage 2 Examination practice paper		

UNIT 3

UNIT 3 (cont.)

Maths language in exams

When a question says...	What it means
You must show your working...	You will lose marks if you do not show how you worked out the answer.
Estimate...	Usually means round numbers to 1 significant figure and then carry out the calculation.
Calculate...	Some working out is needed – so show it!
Work out OR Find...	A written or mental calculation is needed.
Write down...	Written working out is not usually required.
Give an exact value of...	No rounding or approximations.
Give your answer to an appropriate degree of accuracy...	If the numbers in the question are given to 2 decimal places, give your answer to 2 decimal places.
Give your answer in its simplest form...	Usually means you will need to cancel a fraction or a ratio.
Simplify...	In algebra, means collect like terms together.
Solve...	Usually means find the value of x in an equation.
Expand...	Multiply out the brackets.
Factorise...	Put in brackets with common factors outside the bracket.
Measure...	Use a ruler or a protractor to measure lengths or angles accurately.
Draw an accurate diagram...	Use a ruler and protractor to draw the diagram. Lengths must be exact and angles must be accurate.
Construct, using ruler and compasses...	Draw, using a ruler as a straight edge and compasses to draw arcs. Leave your construction lines and arcs in – don't rub them out.
Sketch...	An accurate drawing is not required; a freehand drawing will be fine.
Diagram NOT accurately drawn...	Don't measure angles or sides. If you are asked to find them, you need to work them out.
Give reasons for your answer OR Explain why...	You need to write an explanation. Show any working out, or quote any laws or theorems you used, for example Pythagoras' theorem.
Use your (the) graph...	Read values from your graph and use them.
Describe fully...	Usually means transformations: • Reflection – give the equation of the line of reflection (2 marks) • Rotation – give the angle, direction of turn and the centre of rotation (3 marks) • Enlargement – give the scale factor and the centre of enlargement (3 marks)
Give a reason for your answer...	In angle questions, means write a reason. For example: • angles in a triangle add up to 180° • alternate angles

Collecting data

- Data which can be counted is called **discrete data**. For example, the number of cars in a car park.

- Data which is measured is called **continuous data**. For example, height in centimetres or weight in kilograms.

- A **tally chart** is a way of recording and displaying data.

Key words

| data ☐ | continuous ☐ |
| discrete ☐ | tally chart ☐ |

Example

Grade G

20 people bought cakes at a cake stall. They bought

chocolate	fruit	lemon	banana	fruit
fruit	chocolate	chocolate	fruit	chocolate
banana	chocolate	fruit	chocolate	lemon
chocolate	banana	chocolate	chocolate	fruit

(a) Draw a tally chart to show this information.

(b) What special mathematical name is used for this type of data?

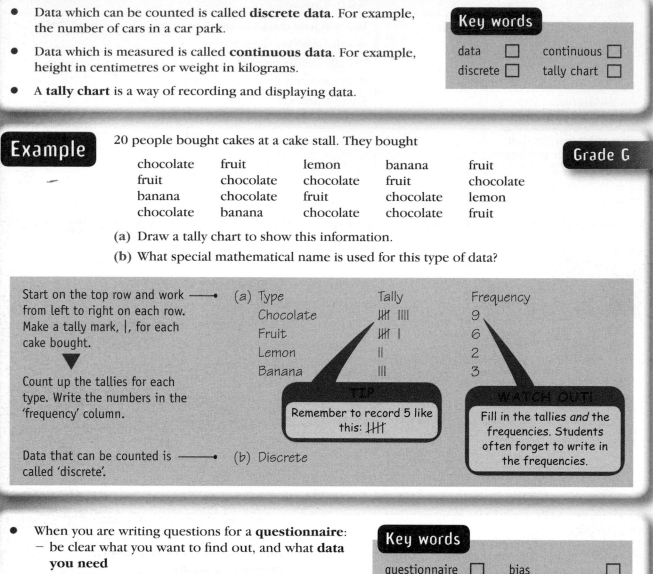

Start on the top row and work from left to right on each row. Make a tally mark, |, for each cake bought.

▼

Count up the tallies for each type. Write the numbers in the 'frequency' column.

(a)
Type	Tally	Frequency
Chocolate	IIII IIII	9
Fruit	IIII I	6
Lemon	II	2
Banana	III	3

TIP
Remember to record 5 like this: IIII

WATCH OUT!
Fill in the tallies *and* the frequencies. Students often forget to write in the frequencies.

Data that can be counted is called 'discrete'. → (b) Discrete

- When you are writing questions for a **questionnaire**:
  – be clear what you want to find out, and what **data you need**
  – ask short, simple questions
  – provide **response boxes** with possible answers
  – avoid questions which are vague, too personal, or which may influence the answer (**leading questions**).

Key words

questionnaire ☐	bias ☐
data ☐	leading question ☐
response box ☐	

Example

A shopkeeper wants to find out how many chocolate bars students eat. He uses this question in his questionnaire: 'You enjoy eating chocolate bars, don't you?'

Grade D

(a) Explain why this is not a good question to ask.

His next question is 'How many chocolate bars have you eaten?' A few ☐   A lot ☐

Grade D

(b) Write down two things that are wrong with this question.

Grade C

(c) Write down an improved question that the shopkeeper could use. Include response boxes.

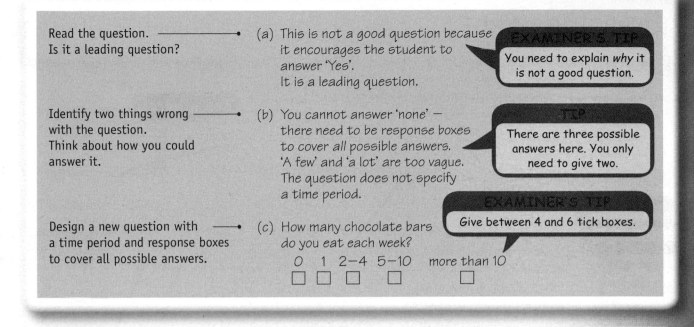

Read the question. ⟶ (a) This is not a good question because it encourages the student to answer 'Yes'.
Is it a leading question?
It is a leading question.

EXAMINER'S TIP
You need to explain *why* it is not a good question.

Identify two things wrong ⟶ (b) You cannot answer 'none' – there need to be response boxes to cover *all* possible answers. 'A few' and 'a lot' are too vague. The question does not specify a time period.
with the question.
Think about how you could answer it.

TIP
There are three possible answers here. You only need to give two.

Design a new question with ⟶ (c) How many chocolate bars do you eat each week?
a time period and response boxes to cover all possible answers.

0   1   2–4   5–10   more than 10
☐   ☐   ☐   ☐   ☐

EXAMINER'S TIP
Give between 4 and 6 tick boxes.

Practice

Grade G

1 Peter carried out a survey to find his friends' favourite juices. Here are his results:

orange	grapefruit	cranberry	tropical
orange	tropical	orange	orange
grapefruit	orange	cranberry	tropical
orange	grapefruit	tropical	orange
cranberry	tropical	cranberry	tropical

(a) Draw a tally chart to show Peter's results.

(b) How many of Peter's friends chose grapefruit as their favourite juice?

(c) Which juice was the most popular?

Grade G

2 Which of the following are discrete data and which are continuous?

(a) the height of a door

(b) the number of words on a page

(c) the number of towns in Essex

(d) the weight of a bag of potatoes

3 The owner of a café uses this question in a questionnaire:

'How much money do you spend in the café?'

A lot ☐   Not much ☐

(a) Write down one thing that is wrong with this question.

(b) Design a better question for the café owner to use. Include response boxes.

Grade D

Grade C

4 Holly wants to carry out a survey about pets. She decides to ask some people whether they prefer dogs, cats, hamsters, rabbits or goldfish.
Design a data collection sheet that she can use to carry out the survey.

Grade C

Check your answers on page 162. For full worked solutions see the CD.
See the Student Book on the CD if you need more help.

Question	1	2	3a	3b	4
Grade	G	G	D	C	C
Student Book pages	U1 4–7	U1 4–7	U1 2–4		U1 4–7

Organising data

- A **database** is an organised collection of **information**.
  It can be stored in a table or on a computer.

Key words

database ☐   information ☐

Example

This table gives some information about holidays in Spain.

Town	Hotel	Children's club	Price	
			8 days	15 days
Mijas	Neptune	No	555	769
Malaya	Don Paco	Yes	435	649
Nerja	Europa	Yes	495	718
Torremolinos	Cavona	No	602	819
Puerto Banus	Marina	No	595	789

Grade G (a) Write down the name of the hotel at Nerja.

Grade F (b) Which hotels have 15-day holidays for less than £750?

Grade F (c) Which hotels do not have a children's club, and have 8-day holidays for between £500 and £600?

Look down the 'town' column to find Nerja. Read across that row to find the hotel name. ————• (a) Europa

Look down the '15 days' column to find prices less than £750. ————• (b) Don Paco, Europa

Find the hotels without a children's club. Then look at the 8-day prices for these hotels. ————• (c) Neptune, Marina

WATCH OUT!
There may be more than one answer. Students often give only one.

TIP
Don't stop when you find one answer. Work through and find *all* the possible answers.

- **Distance charts** give the distances between towns.

- **Two-way tables** are used to record or display information that is grouped in two categories.

Key word

two-way table ☐

Example   **Grade E**

Lucy interviewed 100 people who buy coffee.
She asked them which type they buy most, and in which size packets.
Some of her results are given in the table:
Fill in the blanks.

	Instant	Beans	Ground	Total
50 g	3	1	0	
100 g	16			36
250 g	32	9		
Total		18		100

Look for a row or column with one missing value — cell **1**
Fill in this value, which is the total of the three cells above
3 + 16 + 32 = 51
Now look for other rows or columns with one missing value, and complete the table, filling in the empty cells in turn

	Instant	Beans	Ground	Total
50 g	3	1	0 ³	4
100 g	16 ²	8 ⁴	12	36
250 g	32	9 ⁶	19 ⁵	60
Total	¹ 51	18 ⁷	31	100

TIP
Start with a row or column with only *one* missing value.

TIP
Check that additions work both ways.

Practice

1 This table shows the marks of five students.

Name	English	Maths	Science	History
Stephen	23	29	21	26
Jackie	28	26	29	28
Jason	25	30	23	27
Sakina	26	28	27	32
Mary	32	24	19	25

Grade G

(a) Which student had the highest English mark?

Grade G

(b) Which students had science marks higher than 25?

Grade F

(c) Which student got 28 marks in two subjects?

Grade E

2 Class 11S asked 80 adults which type of television programme they enjoyed most. The two-way table shows information about some of their answers.

	Comedy	Soap	Documentary	News	Total
Men	7			9	
Women		23	6	2	
Total	19		16		80

Complete the table.

Check your answers on pages 162. For full worked solutions see the CD.
See the Student Book on the CD if you need more help.

Question	1ab	1c	2
Grade	G	F	E
Student Book pages	U1 40–42		U1 7–10

Collecting and organising data: topic test

Check how well you know this topic by answering these questions.
First cover the answers on the facing page.

Test questions

1 A teacher asked a class of 20 students how they got to school. They said

| walk | car | bike | bus | train | car | car | walk | train | car |
| train | walk | car | bus | car | car | bus | bus | car | car |

Show this information in a tally chart.

2 This table gives information about cordless electric drills.

Make	Volts	Speed	Torque	Number of gears
Challenger	14.4	700	16	1
Block & Ducker	18	850	5	1
Worker	14.4	1300	24	2
Boss	24	1150	16	1
De Wizz	12	1400	17	2

(a) Which make of drill has a torque value of 24? **(b)** Which makes of drill have two gears?

(c) Which makes of drill have speeds between 800 and 1200?

3 Here is a mileage chart:

(a) Use the mileage chart to find the distance between Bristol and Plymouth.

(b) Which two towns are closest together?

(c) Joe travels from Barnstaple to Exeter, then from Exeter to Plymouth, and then from Plymouth to Penzance. How far does he travel altogether?

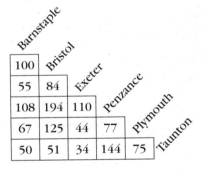

4 100 students each study French, Italian or German.
Complete this two-way table, which shows some information about the students.

	French	Italian	German	Total
Boys	21	20		46
Girls	17			
Total			29	100

5 Maddy wants to carry out a survey into how much time people spending listening to the radio.
Here is part of her questionnaire:

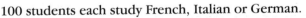

You listen to the radio, don't you? How much?

Sometimes ☐ Always ☐

(a) Write down two things that are wrong with this question.

(b) Write down an improved question that Maddy could use. Include response boxes.

Now check your answers – see the facing page.

Cover this page while you answer the test questions opposite.

Worked answers

Revise this on...

G 1

Travel method	Tally	Total
Walk	III	3
Car	IIII IIII	9
Bike	I	1
Bus	IIII	4
Train	III	3

page 2

G 2 (a) Worker
   (b) Worker, De Wizz
   (c) Block & Ducker, Boss

page 4

G 3 (a) 125 miles

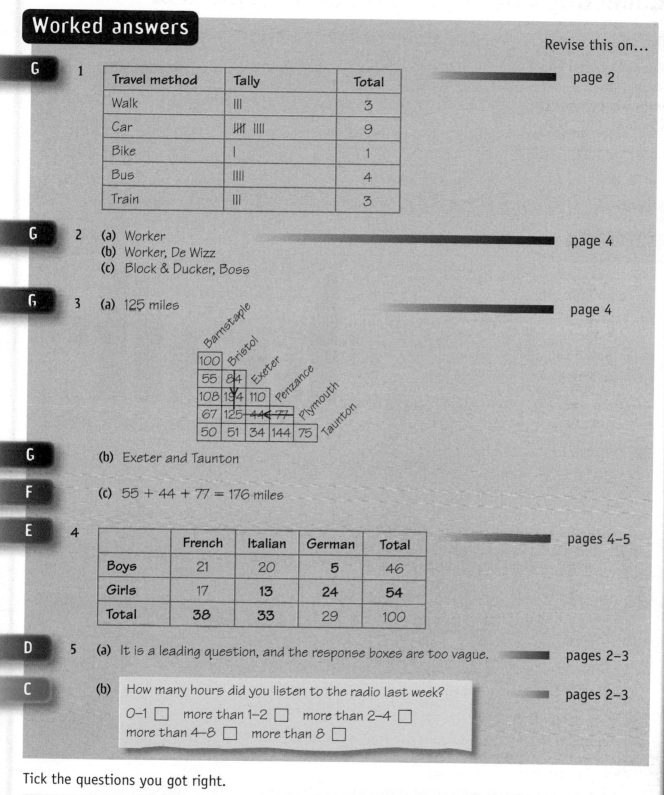

page 4

G   (b) Exeter and Taunton

F   (c) 55 + 44 + 77 = 176 miles

E 4

	French	Italian	German	Total
Boys	21	20	5	46
Girls	17	13	24	54
Total	38	33	29	100

pages 4–5

D 5 (a) It is a leading question, and the response boxes are too vague.

pages 2–3

C   (b) How many hours did you listen to the radio last week?

0–1 ☐   more than 1–2 ☐   more than 2–4 ☐
more than 4–8 ☐   more than 8 ☐

pages 2–3

Tick the questions you got right.

Question	1	2	3ab	3c	4	5a	5b
Grade	G	G	G	F	E	D	C

Mark the grade you are working at on your revision planner on page viii.

Charts

- A **pictogram** uses symbols or pictures to represent quantities.

- It needs a **key** to show what one symbol represents.

- A **bar chart** shows data that can be counted. You must leave a gap between the bars.

- A **dual bar chart** compares two sets of data.

Example

Grade F

This bar chart shows the numbers of cars that Evan and Lindsey sold last week.

(a) How many cars did Evan sell on Monday?

(b) On which day did Lindsey sell 9 cars?

(c) Work out the total number of cars sold on Tuesday.

(d) Who sold the greater number of cars on Friday and Saturday?

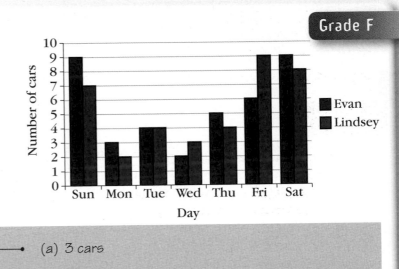

Use the key to see which colour bars show Evan's sales.
Find Evan's bar for Monday and read off the number of sales.

(a) 3 cars

Find 9 on the 'number of cars' axis.
Find a bar for Lindsey that is this height.

(b) Friday

Find the number of cars Evan sold and the number Lindsey sold and add them together.

(c) Evan 4, Lindsey 4
4 + 4 = 8 cars

EXAMINER'S TIP

Show your working

Work out Evan's sales for Friday and Saturday.
Work out Lindsey's sales for Friday and Saturday.
Who sold more?

(d) Evan 6 + 9 = 15
Lindsey 9 + 8 = 17
Lindsey sold more cars.

- A **histogram** shows grouped continuous data.

- A **frequency polygon** shows the general pattern of data represented by a histogram.

Example

Grade C

The table shows the times, in minutes, it takes some people to finish a crossword.

Draw a frequency polygon for this data.

Time, t (minutes)	Frequency
$0 \leqslant t < 5$	4
$5 \leqslant t < 10$	8
$10 \leqslant t < 15$	13
$15 \leqslant t < 20$	16
$20 \leqslant t < 25$	6
$25 \leqslant t < 30$	3

Draw the histogram.
▼
Mark the mid-point of each bar.
▼
Join the mid-points of the bars for the frequency polygon.

EXAMINER'S TIP

You could just plot the mid-points of the bars to draw the frequency polygon.

Practice

1 The pictogram shows the numbers of teas sold in George's café.

Grade G

(a) Write down the number of teas sold on
(i) Monday (ii) Wednesday.

(b) 10 teas were sold on Thursday, and 35 teas on Friday. Complete the pictogram.

Monday	▱ ▱ ▱
Tuesday	▱ ▱ ▱ ▱
Wednesday	▱ ▱ ▯
Thursday	
Friday	

Key: ▱ represents 10 teas

2 Here is a dual bar chart to show Stephen's and Lucy's marks in four tests.

Grade F

(a) What was Lucy's score in English?

(b) In which subject did Stephen score 15?

(c) In which subject did they both score the same?

(d) What was Lucy's total score in all the tests?

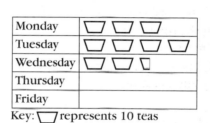

3 The table shows some people's weights in kilograms. Draw a frequency polygon for this data.

Grade C

Weight, w (kilograms)	Frequency
$50 \leqslant w < 55$	2
$55 \leqslant w < 60$	5
$60 \leqslant w < 65$	11
$65 \leqslant w < 70$	13
$70 \leqslant w < 75$	7
$75 \leqslant w < 80$	2

Check your answers on page 162. For full worked solutions see the CD.

See the Student Book on the CD if you need more help.

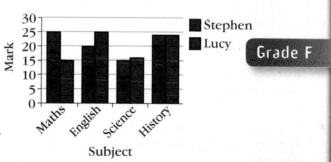

Question	1	2	3
Grade	G	F	C
Student Book pages	U1 13–16	U1 42–45	U1 21–23

Pie charts and stem and leaf diagrams

- A **pie chart** is a way of displaying data when you want to show how something is shared or divided.

- The angles at the centre of a pie chart add up to 360°.

Example   Grade F/E

Leonie asked 180 Year 9 students 'What is your favourite drink?'

Her results are shown in this pie chart.

Work out the number of students who preferred each drink.

Measure each angle using a protractor. ⟶
Tea: 60°   Coffee: 80°
Fizzy: 130°   Water: 90°

TIP
Check that the angles add up to 360°.

Work out how many degrees ⟶ represent one student.
180 students = 360°
So 1 student $= \frac{360°}{180} = 2°$

Work out the number of ⟶ students for each angle.
Tea: the angle is 60°
so the number of students is $\frac{60°}{2°} = 30$

WATCH OUT!
Remember to *divide* by the angle for each student. Students often multiply.

Coffee: number of students is $\frac{80°}{2°} = 40$

Fizzy: number of students is $\frac{130°}{2°} = 65$

Water: number of students is $\frac{90°}{2°} = 45$

Example   Grade E

Ishmael asked some people their favourite colour.

The table shows his results.

Draw an accurate pie chart to show this information.

Colour	Number of people
Red	50
Blue	90
Green	30
Black	70

Work out the total number of ⟶ people.
$50 + 90 + 30 + 70 = 240$

TIP
240 people are represented by 360° (the angle at the centre).

Work out the angle for one person.
1 person is represented by $\frac{360°}{240} = 1.5°$

Work out the angle for each ⟶ colour.
Multiply the number of people by the angle for one person.
Red: $50 \times 1.5° = 75°$
Blue: $90 \times 1.5° = 135°$
Green: $30 \times 1.5° = 45°$
Black: $70 \times 1.5° = 105°$

EXAMINER'S TIP
Always show your working.

TIP
Check that the angles add up to 360°.

Draw the pie chart carefully, ⟶ using a protractor.

TIP
Remember to label each sector on your pie chart.

- A **stem and leaf diagram** shows the shape of a distribution and keeps all the data values.

- It needs a **key** to show how the stem and leaf are combined.

Example

Grade D

Jane throws a dart 20 times. Here are her scores:

| 15 | 30 | 23 | 9 | 37 | 42 | 49 | 36 | 49 | 55 |
| 66 | 57 | 69 | 62 | 38 | 31 | 20 | 46 | 17 | 37 |

Draw an ordered stem and leaf diagram to show these scores. Include a key.

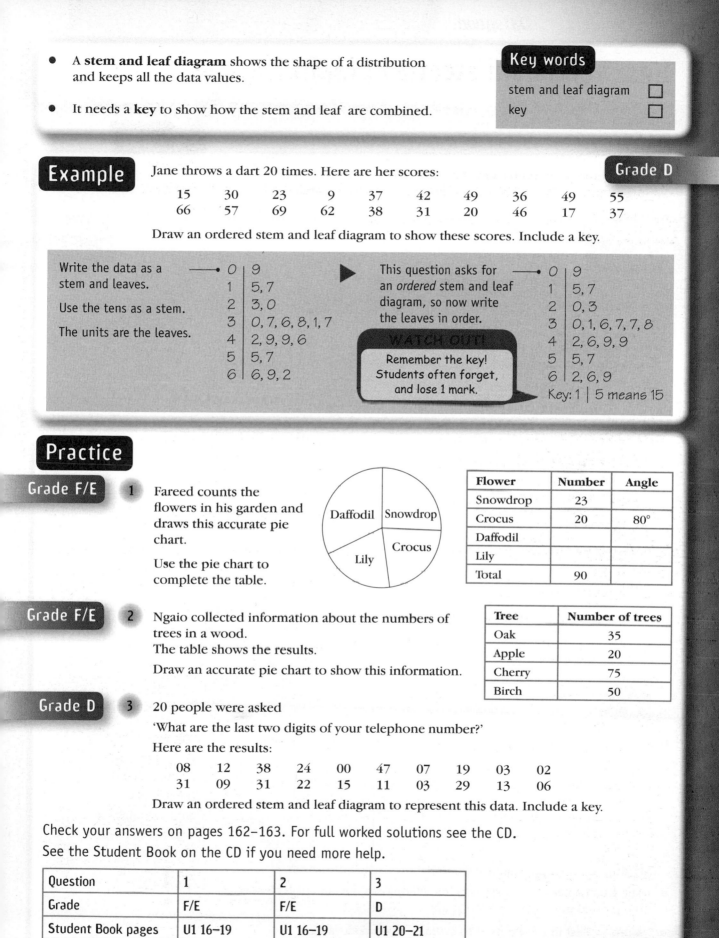

Write the data as a stem and leaves.

Use the tens as a stem.

The units are the leaves.

```
0 | 9
1 | 5, 7
2 | 3, 0
3 | 0, 7, 6, 8, 1, 7
4 | 2, 9, 9, 6
5 | 5, 7
6 | 6, 9, 2
```

▶ This question asks for an *ordered* stem and leaf diagram, so now write the leaves in order.

WATCH OUT!
Remember the key! Students often forget, and lose 1 mark.

```
0 | 9
1 | 5, 7
2 | 0, 3
3 | 0, 1, 6, 7, 7, 8
4 | 2, 6, 9, 9
5 | 5, 7
6 | 2, 6, 9
```
Key: 1 | 5 means 15

Practice

Grade F/E **1** Fareed counts the flowers in his garden and draws this accurate pie chart.

Use the pie chart to complete the table.

Flower	Number	Angle
Snowdrop	23	
Crocus	20	80°
Daffodil		
Lily		
Total	90	

Grade F/E **2** Ngaio collected information about the numbers of trees in a wood.
The table shows the results.

Draw an accurate pie chart to show this information.

Tree	Number of trees
Oak	35
Apple	20
Cherry	75
Birch	50

Grade D **3** 20 people were asked

'What are the last two digits of your telephone number?'

Here are the results:

| 08 | 12 | 38 | 24 | 00 | 47 | 07 | 19 | 03 | 02 |
| 31 | 09 | 31 | 22 | 15 | 11 | 03 | 29 | 13 | 06 |

Draw an ordered stem and leaf diagram to represent this data. Include a key.

Check your answers on pages 162–163. For full worked solutions see the CD.
See the Student Book on the CD if you need more help.

Question	1	2	3
Grade	F/E	F/E	D
Student Book pages	U1 16–19	U1 16–19	U1 20–21

Time series and scatter graphs

- A **line graph** can be used to show continuous data.

- A line graph used to illustrate data collected at intervals in time (e.g. hourly, daily, weekly, …) is called a **times series graph**.

Example

The graph shows the percentage of trains that arrived on time each month from January to August.

(a) From the graph, write down the percentage of trains that arrived on time in May.

(b) In which month did the highest percentage of trains arrive on time? What was the percentage?

(c) The percentages of trains arriving on time for the rest of the year were

September 90%   October 86%
November 75%   December 89%

Complete the graph for these months.

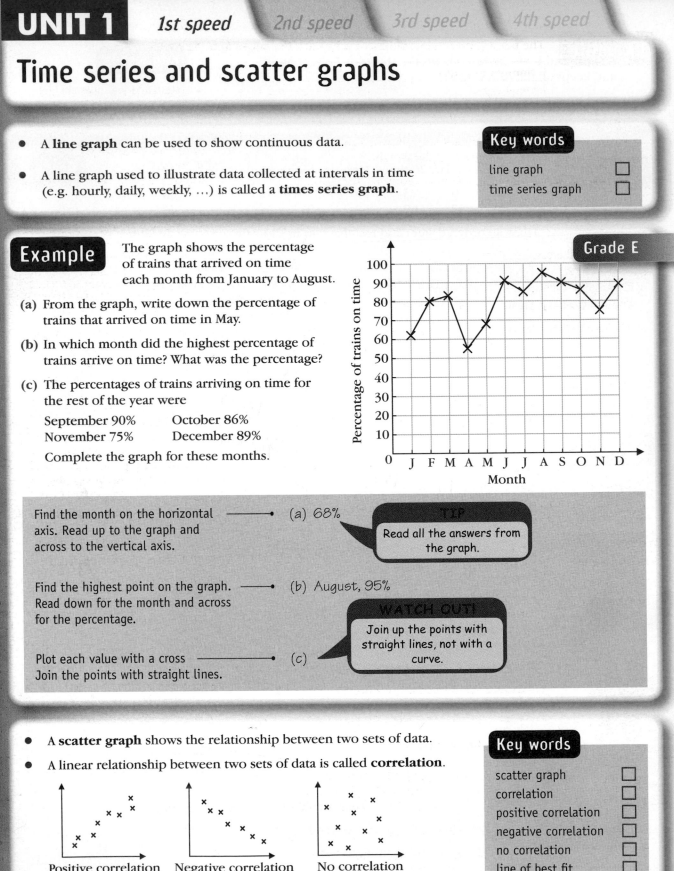

Grade E

Find the month on the horizontal axis. Read up to the graph and across to the vertical axis. ——• (a) 68%

TIP
Read all the answers from the graph.

Find the highest point on the graph. Read down for the month and across for the percentage. ——• (b) August, 95%

WATCH OUT!
Join up the points with straight lines, not with a curve.

Plot each value with a cross ——• (c)
Join the points with straight lines.

- A **scatter graph** shows the relationship between two sets of data.

- A linear relationship between two sets of data is called **correlation**.

Positive correlation   Negative correlation   No correlation

Key words
scatter graph ☐
correlation ☐
positive correlation ☐
negative correlation ☐
no correlation ☐
line of best fit ☐

- The **line of best fit** is a straight line that passes through or is close to the plotted points on a scatter graph.

- A line of best fit can be used to estimate other data values.

Example

The table shows the numbers of pages in nine books and their weights in grams.

Number of pages	65	115	85	125	100	75	145	125	90
Weight (g)	150	260	170	280	220	170	310	260	200

Grade D (a) Draw a scatter graph to represent this data.

Grade D (b) Describe the relationship between the number of pages and the weight.

Grade C (c) Draw a line of best fit on your scatter graph.

(d) Use your line of best fit to estimate **Grade C**

(i) the number of pages in a book of weight 265 g

(ii) the weight of a book with 110 pages.

Plot each pair of values on the scatter graph with a cross. ———————— (a)

Identify the type of correlation. ——— (b) As the number of pages increases, the weight increases. It is positive correlation.

Draw a straight line as close to as many of the points as possible. ——— (c) The line of best fit is the solid red line on the graph.

Draw lines across and ——— (d) (i) 121 pages
down, and read off the (ii) 242 g
values.

EXAMINER'S TIP
Draw lines on the scatter graph to get accurate readings (to the nearest half-square).

TIP
When drawing a line of best fit it is best to use a clear plastic ruler. There should be roughly equal numbers of points above and below the line.

Practice

1 The table shows the temperatures, in °C, from 08 00 to 12 00 one day.

Grade E

(a) Draw a time series graph for this information.

(b) Estimate a temperature for 11 30.

(c) Describe the general trend in the temperature.

Time	Temperature (°C)
08 00	2
09 00	3
10 00	5
11 00	9
12 00	15

2 The table shows the temperature recorded by a weather balloon at different heights.

Grade D

(a) Draw a scatter graph to represent this data.

Grade D

(b) What type of correlation do you find?

Grade C

(c) Draw a line of best fit.

Grade C

(d) Use your scatter graph to estimate

(i) the temperature at a height of 2.4 km

(ii) the height where a temperature of 20 °C might be recorded.

Height (km)	Temperature (°C)
0.4	23
1	17
1.6	12
2	10
2.7	6
3	5

Check your answers on page 163. For full worked solutions see the CD.
See the Student Book on the CD if you need more help.

Question	1	2ab	2cd
Grade	E	D	C
Student Book pages	U1 54–58	U1 51–54	

13

Presenting data: topic test

Check how well you know this topic by answering these questions.
First cover the answers on the facing page.

Test questions

1 Jez has a stamp collection. The pictogram shows the numbers of stamps he has from France, Germany and Italy.

(a) Write down the number of stamps from France.

(b) Write down the number of stamps from Germany.

Jez has 60 stamps from Spain and 30 stamps from Austria.

(c) Use this information to complete the pictogram.

France	⊞ ⊞ ⊞ ⊞
Germany	⊞ ⊞ ⊟
Italy	⊞ ⊞ ⊞ ⊞
Spain	
Austria	

Key: ⊞ represents 20 stamps

2 This accurate pie chart shows information about the medals won by the UK in the Paralympic Games in Athens 2004.

The total number of medals won by the UK was 94.

(a) Find the number of gold, silver and bronze medals.

(b) What fraction of the total medals won were silver?

3 An internet company recorded the number of orders it received on each of 30 days.
Here are the results:

18	48	35	12	43	26	40	14	26	16
26	13	58	39	36	13	38	57	16	38
29	44	29	26	44	26	51	52	24	15

Represent this data using an ordered stem and leaf diagram. Include a key.

4 The table shows the test marks for eight students.

(a) Draw a scatter graph to show this information.

(b) Describe the relationship between the two sets of data.

(c) Draw a line of best fit on your scatter graph.

(d) Fatima's maths mark was 15.
Use your line of best fit to estimate her science mark.

Maths	25	6	17	33	21	10	17	28
Science	20	8	15	29	22	9	19	30

5 Toby measures the weights of the tomatoes from his tomato plants.
His results are summarised in the table.

(a) On one sets of axes, draw two frequency polygons to show this information.

(b) Comment on Toby's plants.

Weight (g)	Frequency plant A	Frequency plant B
30−39	1	3
40−49	2	7
50−59	3	9
60−69	6	7
70−79	5	4
80−89	3	2

Now check your answers – see the facing page.

Cover this page while you answer the test questions opposite.

Worked answers

Revise this on...

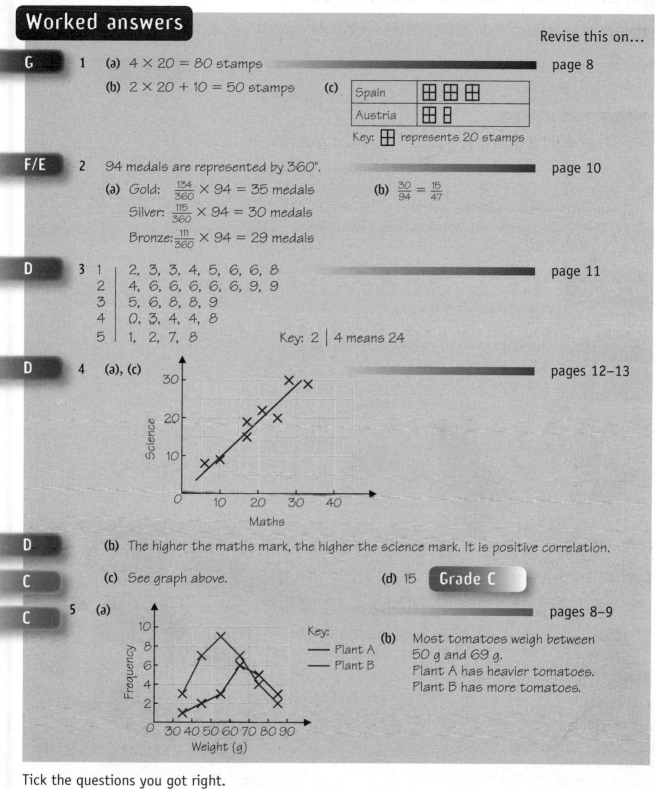

G 1 (a) $4 \times 20 = 80$ stamps — page 8

(b) $2 \times 20 + 10 = 50$ stamps (c)

| Spain | ⊞ ⊞ ⊞ |
| Austria | ⊞ ⧄ |

Key: ⊞ represents 20 stamps

F/E 2 94 medals are represented by 360°. — page 10

(a) Gold: $\frac{134}{360} \times 94 = 35$ medals (b) $\frac{30}{94} = \frac{15}{47}$

Silver: $\frac{115}{360} \times 94 = 30$ medals

Bronze: $\frac{111}{360} \times 94 = 29$ medals

D 3
1	2, 3, 3, 4, 5, 6, 6, 8
2	4, 6, 6, 6, 6, 6, 9, 9
3	5, 6, 8, 8, 9
4	0, 3, 4, 4, 8
5	1, 2, 7, 8

— page 11

Key: 2 | 4 means 24

D 4 (a), (c) — pages 12–13

D (b) The higher the maths mark, the higher the science mark. It is positive correlation.

C (c) See graph above. (d) 15 **Grade C**

C 5 (a) — pages 8–9

Key:
— Plant A
— Plant B

(b) Most tomatoes weigh between 50 g and 69 g.
Plant A has heavier tomatoes.
Plant B has more tomatoes.

Tick the questions you got right.

Question	1	2	3	4ab	4cd	5
Grade	G	F/E	D	D	C	C

Mark the grade you are working at on your revision planner on page viii.

15

Averages and the range (I)

- The **mode** of a set of data is the value which occurs most often.

- The **median** is the middle value when the data are arranged in order of size.

- The **mean** of a set of data is the sum of the values divided by the number of values.

- The **range** of a set of data is the difference between the highest value and the lowest value.

Key words

mode ☐
median ☐
mean ☐
range ☐

Example   Murray made this list of his test marks:

4, 2, 6, 6, 6, 5, 4, 2, 1

Grade G   (a) Write down the mode of his test marks.

Grade G   (b) Work out the median of his test marks.

Grade F   (c) Work out the mean of his test marks.

Grade F   (d) Work out the range of his test marks.

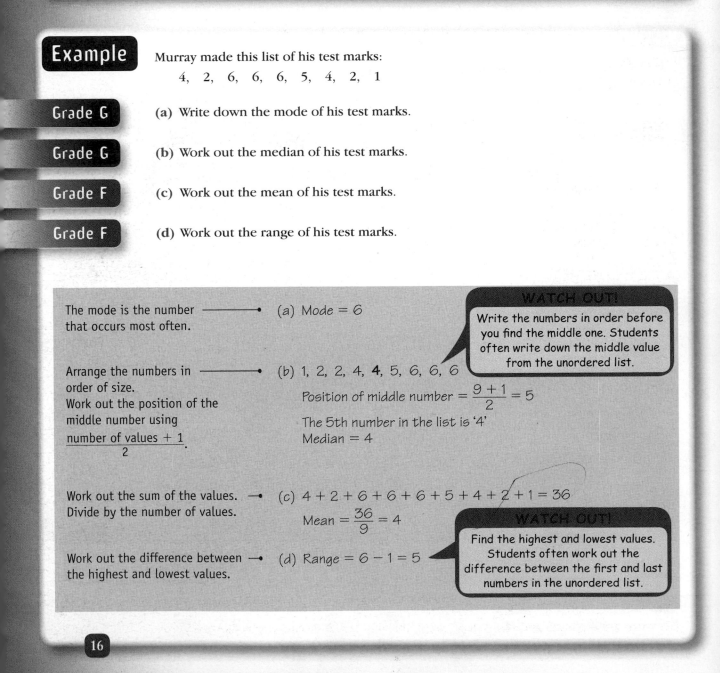

The mode is the number that occurs most often. → (a) Mode = 6

WATCH OUT! Write the numbers in order before you find the middle one. Students often write down the middle value from the unordered list.

Arrange the numbers in order of size. Work out the position of the middle number using $\frac{\text{number of values} + 1}{2}$.

(b) 1, 2, 2, 4, **4**, 5, 6, 6, 6

Position of middle number = $\frac{9+1}{2} = 5$

The 5th number in the list is '4'
Median = 4

Work out the sum of the values. → (c) 4 + 2 + 6 + 6 + 6 + 5 + 4 + 2 + 1 = 36
Divide by the number of values.

Mean = $\frac{36}{9} = 4$

WATCH OUT! Find the highest and lowest values. Students often work out the difference between the first and last numbers in the unordered list.

Work out the difference between the highest and lowest values. → (d) Range = 6 − 1 = 5

- With a **frequency table**:

$$\text{mean} = \frac{\Sigma fx}{\Sigma f}$$

the sum of all the ($f \times x$) values in the table

the sum of the frequencies

Key words

frequency table ☐

Example

Luxni has some boxes of candles.

The table gives information about the numbers of candles in each box.

Grade F   **(a)** How many boxes does Luxni have?

Grade F   **(b)** Write down the modal number of candles in a box.

Grade E   **(c)** Work out the median number of candles in a box.

Grade D   **(d)** Work out the mean number of candles in a box.

Number of candles	Frequency
3	4
4	10
5	5
6	1

Add the numbers in the frequency column.

(a) Number of boxes = 4 + 10 + 5 + 1 = 20

The mode is the one with the highest frequency.

(b) Mode = 4

WATCH OUT!

Remember to write down the *value*. Students often write down the frequency.

The median is the middle value of the data.

(c) Position of middle value = $\frac{20 + 1}{2}$ = 10.5

Boxes 1–4 have 3 candles.
Boxes 5–14 have 4 candles.
So boxes 10 and 11 each have 4 candles.
Median = $\frac{4 + 4}{2}$ = 4

TIP

The number of candles in 10 boxes of 4 is 10 × 4

Add a third column to the table – work out frequency × number for each row.

(d)

Number of candles x	Frequency f	Frequency × number of candles f × x
3	4	12
4	10	40
5	5	25
6	1	6
Total	20	83

Work out the mean.

Mean = total number of candles ÷ total number of boxes
= 83 ÷ 20
= 4.15 candles

TIP

This is the total number of candles.

For more on averages and the range, including practice questions, see pages 18–19.

Averages and the range (II)

- For **grouped data**:
  - the **modal class** is the class interval with the highest frequency
  - you can state the **class interval** that contains the median
  - you can calculate an estimate of the mean using the middle value of each class interval.

Key words

grouped data ☐
class interval ☐
modal class ☐

Example

The table gives information about the weights of 40 small children.

Grade D

(a) Write down the modal class.

Grade C

(b) Write down the class interval in which the median lies.

Grade C

(c) Work out an estimate for the mean weight.

Weight, w (kg)	Frequency f
$0 < w \leqslant 4$	5
$4 < w \leqslant 8$	13
$8 < w \leqslant 12$	14
$12 < w \leqslant 16$	8

Find the class interval with the highest frequency. ——→ (a) $8 < w \leqslant 12$

Work out the position of the median. ——→ (b) Position of median
$$= \frac{40 + 1}{2} = 20.5$$

▼

Find the class interval that contains the 20th and 21st values. ——→ Class interval of median is $8 < w \leqslant 12$

TIP
$8 < w \leqslant 12$ has the 19th to 32nd values.

Add two more columns to the table — work out the middle value, x, for each row and $f \times x$ for each row. ——→

(c)

Weight, w (kg)	Frequency f	Middle value x	$f \times x$
$0 < w \leqslant 4$	5	2	10
$4 < w \leqslant 8$	13	6	78
$8 < w \leqslant 12$	14	10	140
$12 < w \leqslant 16$	8	14	112
Total	40	Total	340

▼

TIP
Add a row for the totals.

Work out the estimate of the mean. ——→ Estimate of mean
$$= \frac{\text{sum of (middle values} \times \text{frequencies)}}{\text{sum of frequencies}}$$
$$= \frac{340}{40} = 8.5 \text{ kg}$$

WATCH OUT!
Remember to use the *middle* values. Students often use the beginning or end of the class intervals.

Practice

1 Here are the ages, in years, of seven people:

> 37, 28, 33, 33, 29, 30, 27

(a) Write down the mode.

Grade G

(b) Find the median age.

Grade G

(c) Work out the range of the ages.

Grade F

(d) Work out the mean age.

Grade F

2

Number of tries	Number of matches
0	4
1	6
2	11
3	8
4	5

The table gives information about the number of tries scored by a rugby team in each match of the season.

(a) How many matches were there?

Grade F

(b) Write down the modal number of tries in a match.

Grade F

(c) Work out the median number of tries in a match.

Grade E

(d) Work out the mean number of tries in a match.

Grade D

(e) Peter said 'The team scored an average of 5 tries per match.' Explain why this is wrong.

Grade D

3

Time, t (minutes)	Frequency
$10 \leqslant t < 15$	3
$15 \leqslant t < 20$	9
$20 \leqslant t < 25$	18
$25 \leqslant t < 30$	15
$30 \leqslant t < 35$	5

Bronwen recorded the times, in minutes, it took her to complete 50 crosswords.

(a) Write down the modal class.

Grade D

(b) Write down the class interval in which the median lies.

Grade C

(c) Calculate an estimate of the mean time it took Bronwen to complete a crossword.

Grade C

Check your answers on page 163. For full worked solutions see the CD.
See the Student Book on the CD if you need more help.

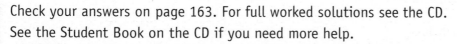

Question	1ab	1cd	2ab	2c	2de	3a	3bc
Grade	G	F	F	E	D	D	C
Student Book pages	U1 27–29			U1 29–31		U1 32–35	

Averages and the range: topic test

Check how well you know this topic by answering these questions.
First cover the answers on the facing page.

Test questions

1 A radio company records the numbers of complaints about its programmes.

Here are the numbers of complaints about a particular programme on each of nine days:

  2, 4, 2, 1, 16, 2, 4, 3, 2

(a) Write down the mode.   (b) Find the median.

(c) Work out the range.   (d) Work out the mean.

2 Mary collects money for a charity.
The table shows the numbers and types of coins that Mary collected last Saturday.

Type of coin	1p	2p	5p	10p	20p	50p	£1	£2
Number of coins	18	35	7	13	9	11	5	1

(a) How many coins did Mary collect?

(b) Write down the modal type of coin that was collected.

(c) Work out the median type of coin.

(d) Calculate the mean value of the coins.

3 In a survey, 50 people were asked how long they spent watching television last weekend.
The results are summarised in the table.

Time, t (hours)	Frequency
$0 \leqslant t < 2$	11
$2 \leqslant t < 4$	15
$4 \leqslant t < 6$	18
$6 \leqslant t < 8$	6

(a) Write down the modal class.

(b) Write down the class interval that contains the median time.

(c) Calculate an estimate of the mean time.

Now check your answers – see the facing page.

Cover this page while you answer the test questions opposite.

Worked answers

Revise this on...

G 1 (a) Mode = 2 — page 16

G (b) Rewrite the numbers in order of size: 1, 2, 2, 2, **2**, 3, 4, 4, 16 — page 16
Median = 2

F (c) Range = highest value − lowest value = 16 − 1 = 15 — page 16

F (d) Mean = $\dfrac{1+2+2+2+2+3+4+4+16}{9} = \dfrac{36}{9} = 4$ — page 16

F 2 (a) 18 + 35 + 7 + 13 + 9 + 11 + 5 + 1 = 99 coins — page 17

F (b) Modal coin = 2p

E (c) Position of median = $\dfrac{99+1}{2} = 50$
Coins 1–18 are 1p coins; coins 19–53 are 2p coins; ...
So coin 50 is a 2p coin.
Median = 2p

D (d) Mean = $\dfrac{\text{sum of (values} \times \text{frequencies)}}{\text{sum of frequencies}}$

$= \dfrac{\Sigma f \times x}{\Sigma f} = \dfrac{1683}{99} = 17p$

Coin	Frequency f	f × x
1p	18	18
2p	35	70
5p	7	35
10p	13	130
20p	9	180
50p	11	550
£1 = 100p	5	500
£2 = 200p	1	200
Totals	99	1683

D 3 (a) Modal class is $4 \leqslant t < 6$ — page 18

C (b) Position of median = $\dfrac{50+1}{2} = 25.5$

The median is the mean of the 25th and 26th times.
The median time lies within the $4 \leqslant t < 6$ class interval.

C (c) Mean = $\dfrac{\Sigma f \times x}{\Sigma f} = \dfrac{188}{50} = 3.76$ hours

Time, t (hours)	Frequency f	Middle value, x	f × x
$0 \leqslant t < 2$	11	1	11
$2 \leqslant t < 4$	15	3	45
$4 \leqslant t < 6$	18	5	90
$6 \leqslant t < 8$	6	7	42
Total	50	Total	188

Tick the questions you got right.

Question	1ab	1cd	2ab	2c	2d	3a	3bc
Grade	G	F	F	E	D	D	C

Mark the grade you are working at on your revision planner on page viii.

Probability (I)

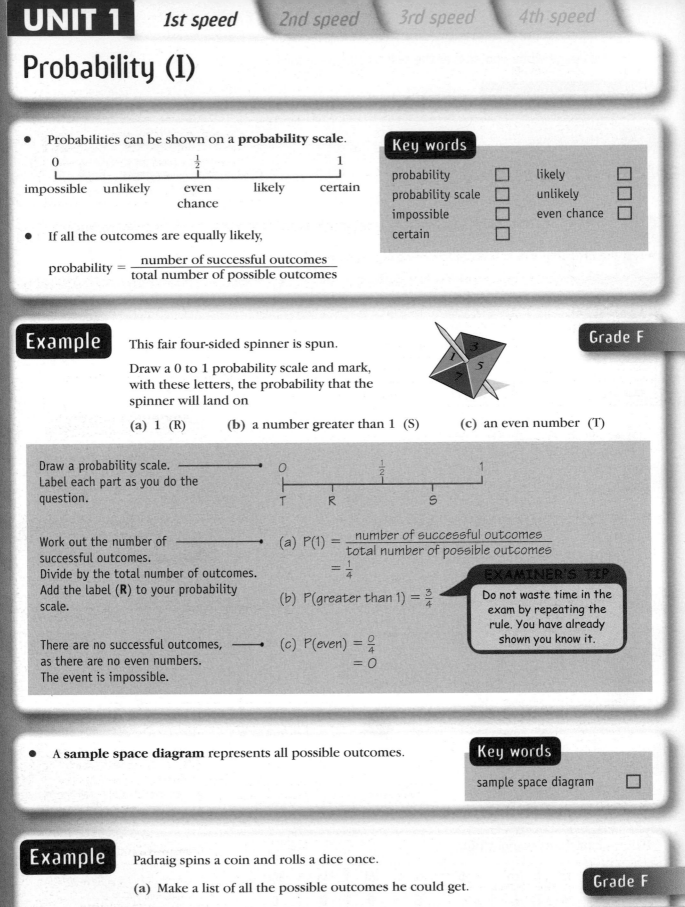

- Probabilities can be shown on a **probability scale**.

0             $\frac{1}{2}$ 1

impossible   unlikely   even   likely   certain
                 chance

Key words

probability	☐	likely	☐
probability scale	☐	unlikely	☐
impossible	☐	even chance	☐
certain	☐		

- If all the outcomes are equally likely,

$$\text{probability} = \frac{\text{number of successful outcomes}}{\text{total number of possible outcomes}}$$

Example   This fair four-sided spinner is spun.

Draw a 0 to 1 probability scale and mark, with these letters, the probability that the spinner will land on

(a) 1 (R)   **(b)** a number greater than 1 (S)   **(c)** an even number (T)

Grade F

Draw a probability scale.
Label each part as you do the question.

O $\frac{1}{2}$ 1

T   R   S

Work out the number of successful outcomes.
Divide by the total number of outcomes.
Add the label (**R**) to your probability scale.

(a) $P(1) = \dfrac{\text{number of successful outcomes}}{\text{total number of possible outcomes}}$
       $= \frac{1}{4}$

(b) $P(\text{greater than } 1) = \frac{3}{4}$

EXAMINER'S TIP
Do not waste time in the exam by repeating the rule. You have already shown you know it.

There are no successful outcomes, as there are no even numbers.
The event is impossible.

(c) $P(\text{even}) = \frac{0}{4}$
         $= 0$

- A **sample space diagram** represents all possible outcomes.

Key words

sample space diagram   ☐

Example   Padraig spins a coin and rolls a dice once.

(a) Make a list of all the possible outcomes he could get.

Grade F

(b) Work out the probability that he gets a tail and a 1.

Grade E

Write out all the possible outcomes. ———→ (a) (T, 1), (H, 1), (T, 2), (H, 2) (T, 3), (H, 3), (T, 4), (H, 4) (T, 5), (H, 5), (T, 6), (H, 6)

TIP
Work carefully, following a pattern so that you do not miss any outcomes.

Use the rule
$$\text{probability} = \frac{\text{number of successful outcomes}}{\text{total number of outcomes}}$$
———→ (b) $P(T, 1) = \frac{1}{12}$

EXAMINER'S TIP
Use your list from part (a) to answer part (b). Even if (a) is wrong, you can still get full marks for (b).

- If the probability of an event happening is p, the probability of it *not* happening is $1 - p$.

Example Mrs Coley chooses a drink from a machine. She can choose tea, coffee, chocolate or soup.

The table shows the probabilities that she chooses tea, coffee or soup.

Grade D

Drink	Tea	Coffee	Chocolate	Soup
Probability	0.3	0.4		0.1

Work out the probability that she chooses chocolate.

Add up the probabilities that you know. ———→
$$P(\text{tea}) + P(\text{coffee}) + P(\text{soup})$$
$$= 0.3 + 0.4 + 0.1$$
$$= 0.8$$

TIP
Write this value in the table and check that the probabilities add up to 1.

Subtract from 1. ———→ $P(\text{chocolate}) = 1 - 0.8 = 0.2$

- **Two-way tables** can be used to help solve probability problems.

Key words
two-way table ☐

Example 100 students each chose one town to visit last week.
The two-way table shows some information about the students.

Grade D

	Colchester	York	Chester	Total
Boys			20	52
Girls	13			
Total		41	38	100

(a) Complete the two-way table.
(b) One of these 100 students is picked at random. Write down the probability that the student chose York.

Work out the missing values.
For more on two-way tables see pages 4–5. ———→ (a)

	Colchester	York	Chester	Total
Boys	8	24	20	52
Girls	13	17	18	48
Total	21	41	38	100

TIP
Check that the rows and columns all add up to the total.

Use the rule
$$\text{probability} = \frac{\text{number of successful outcomes}}{\text{total number of outcomes}}$$
———→ (b) $P(\text{York}) = \frac{41}{100}$

TIP
If your number is greater than 1 you have made a mistake.

For more on probability, including practice questions, see pages 24–25.

Probability (II)

- Estimated probability = $\dfrac{\text{number of successful trials}}{\text{total number of trials}}$

- You can use the estimated probability to predict results.

Key words

estimated probability ☐

Grade C

Example

The probability that a biased coin lands 'heads' is 0.7

Henry is going to spin the coin 200 times.

Work out an estimate for the number of times it will land 'heads'.

Rearrange the rule ———————•

estimated probability = $\dfrac{\text{number of successful trials}}{\text{total number of trials}}$

▼

number of successful trials =
  estimated probability × total number of trials

$P(\text{head}) = 0.7$

Estimated number of heads
in 200 trials = 0.7×200
                  $= 140$

WATCH OUT!
Remember to check that your answer is sensible. Students often put the decimal point in the wrong place.

Grade C

Example   A fair dice is rolled. Work out the probability that it will land on 5 or 6.

Work out the probability of ———————• $P(5) = \frac{1}{6}, P(6) = \frac{1}{6}$
each outcome.

▼

Add the probabilities ———————• $P(5 \text{ or } 6) = \frac{1}{6} + \frac{1}{6} = \frac{2}{6} = \frac{1}{3}$
together.

TIP
For more on adding fractions see pages 94–95.

WATCH OUT!
Be careful to add the fractions correctly. Students often write $\frac{1}{6} + \frac{1}{6}$ as $\frac{2}{12}$, which is wrong.

Grade C

Example   A bag contains 1 white, 3 black and 5 blue beads.
Omar selects a bead at random.
What is the probability that the bead he chooses is blue or white?

Work out the total number of ———————•   Total = $5 + 3 + 1 = 9$ beads
beads, and the probability of $P(\text{white}) = \frac{1}{9}, P(\text{blue}) = \frac{5}{9}$
each outcome.

▼

Add the probabilities together. ———————• $P(\text{blue or white}) = \frac{1}{9} + \frac{5}{9} = \frac{6}{9} = \frac{2}{3}$

TIP
Write the fraction in its simplest form. For more on simplifying fractions see page 52.

Practice

1 Draw a 0 to 1 probability scale and mark, with these letters, the probability that

    (a) when you spin a coin it will land 'heads' (H)

    (b) it will never rain in England (R)

    (c) when you roll a dice you will get a 6 (S)

Grade F

2 Monty plays a game of draughts with his friend.
In draughts, games are won, lost or drawn.
The probability that Monty loses the game is 0.25
The probability that Monty draws is 0.4
Work out the probability that Monty wins the game.

Grade D

3 Freddie asked 50 people how they travelled to work.
The table shows this information.

	Car	Walk	Bus	Train	Total
Men	12			7	24
Women	16		2		26
Total		9	5	8	

Grade D

    (a) Copy and complete the two-way table.

    (b) Freddie chooses a person at random. What is the probability that he chooses

        (i) a person that travels to work by train?

        (ii) a man that walks to work?

        (iii) a woman that travels to work by car?

4 Rhian rolls a blue dice and a red dice.

    (a) List all the possible outcomes.

Grade E

    (b) Use your list to find the probability that she gets a
total score of 7.

Grade C

5 A fair spinner is made in the shape of a regular hexagon.
It can land on red, blue or yellow.
Write down the probability that the spinner will land on red.

Grade C

Check your answers on page 163. For full worked solutions see the CD.
See the Student Book on the CD if you need more help.

Question	1	2	3	4a	4b	5
Grade	F	D	D	E	C	C
Student Book pages	U1 66–69	U1 69–71	U1 69–76	U1 71–74		U1 69–71

Probability: topic test

Check how well you know this topic by answering these questions.
First cover the answers on the facing page.

Test questions

1 This fair six-sided dice is rolled.
Draw a 0 to 1 probability scale and mark, with these
letters, the probability that the dice will land on

(a) a 6 (S)

(b) an odd number (T)

(c) a 7 (V)

2 Vijay spins this fair spinner and tosses a fair coin.
Make a list of all the outcomes he could get.
The first is (1, head).

3 A box contains beads which are red, yellow, blue or green.
Helen is going to pick one bead from the box at random.
The table shows the probabilities that the bead she picks is red, yellow or green.

Colour	Red	Yellow	Blue	Green
Probability	0.15	0.23		0.41

Work out the probability that she will pick a blue bead.

4 80 students in Year 10 each study French, German or Spanish.
The table shows some information about these students.

	French	German	Spanish	Total
Girls	17			46
Boys			20	
Total		23	34	80

(a) Complete the table.

(b) One of these 80 students is picked at random.
Write down the probability that the student studies Spanish.

5 A fair spinner is made in the shape of a regular octagon.
It can land on 5 or 10 or 20.
Write down the probability that the spinner will land on 10.

Now check your answers – see the facing page.

Cover this page while you answer the test questions opposite.

Worked answers

Revise this on...

F 1 (a) $P(6) = \frac{1}{6}$

page 22

    (b) $P(\text{odd}) = \frac{3}{6} = \frac{1}{2}$

    (c) $P(7) = 0$

F 2 (1, head),
(1, tail),
(2, head),
(2, tail),
(3, head),
(3, tail),
(5, head), (5, head)
(5, tail), (5, tail)

page 22–23

D 3 $P(\text{not blue}) = P(\text{red}) + P(\text{yellow}) + P(\text{green})$
               $= 0.15 + 0.23 + 0.41$
               $= 0.79$
$P(\text{blue})\quad = 1 - P(\text{not blue})$
               $= 1 - 0.79$
               $= 0.21$

page 23

D 4 (a)

	French	German	Spanish	Total
Girls	17	15	14	46
Boys	6	8	20	34
Total	23	23	34	80

page 23

    (b) $P(\text{Spanish}) = \dfrac{\text{number of successful outcomes}}{\text{total number of outcomes}}$

                   $= \dfrac{34}{80} = \dfrac{17}{40}$

C 5 $P(10) = \dfrac{\text{number of successful outcomes}}{\text{total number of outcomes}}$

page 22

        $= \dfrac{4}{8} = \dfrac{1}{2}$

Tick the questions you got right.

Question	1	2	3	4	5
Grade	F	F	D	D	C

Mark the grade you are working at on your revision planner on page viii.

Handling data: subject test

Exam practice questions

1 Salih asked his friends 'What is your favourite sport?'
Here are his results:

soccer	golf	soccer	rugby	cricket
rugby	soccer	soccer	golf	soccer
soccer	rugby	cricket	rugby	soccer
golf	soccer	rugby	soccer	rugby

(a) Complete the table to summarise Salih's results.

(b) Write down the number of friends whose favourite sport was rugby.

(c) What was the most popular sport among his friends?

Sport	Tally	Frequency
Soccer		
Golf		
Rugby		
Cricket		

2 The pictogram shows some information about the numbers of DVDs rented from a petrol station.

(a) Write down the number of DVDs rented on
   (i) Saturday   (ii) Sunday.

(b) 40 DVDs were rented on Monday, and 30 DVDs on Tuesday. Show this information on the pictogram.

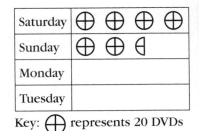

Key: ⊕ represents 20 DVDs

3 Here is a list of Carol's test marks:

   7, 5, 8, 8, 9, 6, 8, 8, 6, 5

(a) Write down the mode.

(b) Work out the mean.

(c) Work out the range.

4 This table gives information about 90 people's eye colour.

Draw an accurate pie chart to show this information.

Eye colour	Number of people
Blue	40
Grey	15
Green	25
Brown	10

5 20 people do a lap round a race track.
Here are their times to the nearest second:

62	46	39	53	28	44	65	41	48	37
36	49	51	46	39	27	60	50	45	33

(a) Draw an ordered stem and leaf diagram to show this information. Include a key.

(b) Use your stem and leaf diagram to write down the median.

6 100 students were asked how they came to school that day.
Some of the results are shown in the two-way table.

	Car	Walk	Cycle	Total
Year 7		13	9	41
Year 8	5			22
Year 9		18		
Total	36		21	100

(a) Complete the two-way table.

(b) One of these students is picked at random.
Write down the probability that this student

(i) came to school by car   (ii) is in Year 7 and cycled to school.

7 A freezer contains four flavours of ice-cream – vanilla, chocolate, strawberry and mint.
The table shows the probabilities that Marcus chooses vanilla, chocolate or mint.

Flavour	Vanilla	Chocolate	Strawberry	Mint
Probability	0.2	0.4		0.15

Work out the probability that he picks a strawberry ice-cream.

8 The table shows the heights and weights of ten students.

Weight (kg)	75	65	82	76	71	65	77	70	72	68
Height (cm)	185	182	191	188	184	166	175	178	181	180

(a) Draw a scatter graph to show this information.

(b) What type of correlation do you find?

(c) Draw a line of best fit.

(d) Use your scatter graph to estimate
(i) the weight of a student whose height is 188 cm
(ii) the height of a student whose weight is 74 kg.

Check your answers on pages 163–164. For full worked solutions see the CD.

Tick the questions you got right.

Question	1	2	3a	3bc	4	5	6a	6b	7	8abc	8d
Grade	G	G	G	F	E	D	E	D	D	D	C
Revise this on page	2	8	16		10	11	23		23	12	

Mark the grade you are working at on your revision planner on page viii.

Go to the pages shown to revise for the ones you got wrong.

Handling data

Collecting and organising data

- A **tally chart** is a way of recording and displaying data.

Flavour	Tally	Frequency
Chocolate	ⅢⅠ ⅠⅠⅠⅠ	9
Fruit	ⅢⅠ Ⅰ	6
Lemon	ⅠⅠ	2
Banana	ⅠⅠⅠ	3

- **Two-way tables** are used to record or display information that is grouped in two categories.

Presenting data

- A **pictogram** uses symbols or pictures to represent quantities. It needs a **key** to show what one symbol represents.

Monday	▭ ▭
Tuesday	▯
Wednesday	▭ ▯
Thursday	▭ ▭ ▭
Friday	▭

Key: ▭ represents 10 bags of toffees

- A **bar chart** shows data that can be counted. You must leave a gap between the bars.

- A **dual bar chart** compares two sets of data.

- A **pie chart** is a way of displaying data when you want to show how something is shared or divided. The angles at the centre of a pie chart add up to 360°.

- A **stem and leaf diagram** shows the shape of a distribution and keeps all the data values. It needs a **key** to show how the stem and leaf are combined.

0	9
1	2, 3
2	0, 7
3	4, 8
4	1

Key: 1 | 2 means 12

- A **line graph** can be used to show continuous data.

Averages and the range

- The **mode** of a set of data is the value which occurs most often.

- The **median** is the middle value when the data are arranged in order of size.

- The **mean** of a set of data is the sum of the values divided by the number of values.

- The **range** of a set of data is the difference between the highest value and the lowest value.

- With a **frequency table**:

$$\text{mean} = \frac{\Sigma fx}{\Sigma f}$$

the sum of all the $(f \times x)$ values in the table

the sum of the frequencies

- For **grouped data**:
  - the **modal class** is the class interval with the highest frequency
  - you can state the **class interval** that contains the median
  - you can calculate an estimate of the **mean** using the middle value of each class interval.

Probability

- Probabilities can be shown on a **probability scale**.

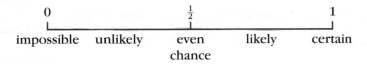

- If all the outcomes are equally likely,

$$\text{probability} = \frac{\text{number of successful outcomes}}{\text{total number of possible outcomes}}$$

- $$\text{Estimated probability} = \frac{\text{number of successful trials}}{\text{total number of trials}}$$

Unit 1 Examination practice paper

A formula sheet can be found on page 161.

Section A (calculator)

1 Jaquii carried out a survey about her friends' pets.

   Here are her results.

Dog	Cat	Rabbit	Hamster	Dog
Cat	Dog	Fish	Cat	Dog
Rabbit	Hamster	Dog	Fish	Fish
Dog	Cat	Fish	Dog	Dog

   (a) Complete the frequency table to show Jaquii's results.

Pet	Tally	Frequency
Dog		
Cat		
Rabbit		
Hamster		
Fish		

   (3 marks)

   (b) Which pet is the most popular? (1 mark)

   (c) Two pets were selected 4 times. Which two? (2 marks)

   (Total 5 marks)

2 These are the number of matches in each of 10 boxes of matches.

   28   28   29   30   30   30   30   30   32   32

   (a) Find the median number of matches in these 10 boxes. (1 mark)

   (b) Work out the range of the number of matches. (2 marks)

   (Total 3 marks)

3 A packet contains seeds that can produce red, yellow, orange or blue flowers.

The table shows the probabilities that a seed taken at randowm from the packet will produce a flower that is red or yellow or orange.

Colour of flower	red	yellow	orange	blue
Probability	0.2	0.3	0.3	

(a) Work out the probability that a seed, taken at random from the packet, will produce a blue flower.

(2 marks)

The packet contains 150 seeds.

(b) Work out the number of seeds that will produce a red flower.

(2 marks)

(Total 4 marks)

4 The table shows information about the 120 tracks on Casey's MP3 player.

Type of music	Number of tracks	Angle
Pop	30	90°
Rock	40	
Classical	15	
Jazz	35	

Complete the pie chart.

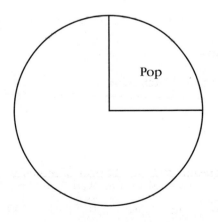

(Total 3 marks)

(Total 15 marks)

Check your answers on page 164. For full worked solutions see the CD.

Section B (non-calculator)

1 The pictogram shows the number of pizzas sold in Mario's restaurant on Monday to Thursday last week.

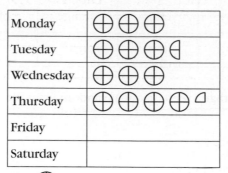

Monday	
Tuesday	
Wednesday	
Thursday	
Friday	
Saturday	

Key: ⊕ represents 4 pizzas

(a) Write down the number of pizzas sold on Monday. **(1 mark)**

(b) Write down the number of pizzas sold on Tuesday. **(1 mark)**

20 pizzas were sold on Friday.
15 pizzas were sold on Saturday.

(c) Use this information to complete the pictogram. **(2 marks)**

(Total 4 marks)

2 The graph shows the number of ice-creams sold each day during one week.

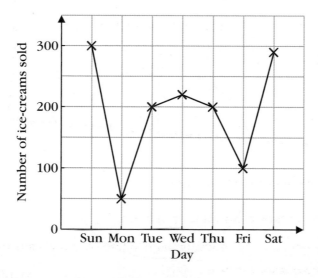

(a) How many ice-creams were sold on Friday? **(1 mark)**

(b) How many more ice-creams were sold on Tuesday than on Monday? **(1 mark)**

(c) Explain what might have happened on Monday. **(1 mark)**

(Total 3 marks)

3 There are 30 students in Tutor Group 11–3.

The two-way table shows some information about what these students eat at lunchtime.

	Sandwiches	School lunch	Lunch at home	Total
Female	5		1	
Male	2	10		14
Total				30

(a) Complete the two-way table. (3 marks)

One of the students is chosen at random.

(b) Find the probability that the student is male and has a school lunch. (2 marks)

(Total 5 marks)

4 Mrs Green makes cakes.

She wants to find out what people think of the cakes she makes.

She uses this question on her questionnaire.

Cake Questionnaire

You do like cakes don't you?

(a) Write down what is wrong with this question. (1 mark)

Mrs Green wants to find out how many cakes people eat.

(b) Design a suitable question for her questionnaire to find out how many cakes people eat.
You must include some response boxes. (2 marks)

(Total 3 marks)

(Total 15 marks)

Check your answers on page 164. For full worked solutions see the CD.

Place value, ordering and rounding

- A number can be written in words or in **figures**.

- Each **digit** in a number has a value that depends on its position. This is its **place value**.

- Digits in a large number are grouped in threes, starting from the *right*, for example 456 762 121

Key words

| figure | ☐ | place value | ☐ |
| digit | ☐ | zero | ☐ |

Example

(a) Write three hundred and forty-six million, five hundred and sixty-one thousand, nine hundred and seventy-eight in figures.

(b) What is the value of the 7 in 37 385?

Grade G

Write the millions. ⟶ (a) 346 000 000   346 millions

TIP
Zeros are used to show that a column is empty.

Write the thousands. ⟶ 561 000   561 thousands

Write the hundreds, tens and units. ⟶ 978   9 hundreds, 7 tens and 8 units

Combine all three parts. ⟶ 346 561 978

Look at the place value column. ⟶ (b) **37** 385
The 7 is in the thousands column.
It means 7 thousands or 7000.

- In whole numbers the more digits, the larger the number.

- When two numbers have the same number of digits, look at the highest place value column to see which is larger. If these digits are the same, look at the next column, and so on.

Example

Write these numbers in order of size, smallest first:

906,   894,   910,   899,   99

Grade G

Find the smallest number first. ⟶ 99   99 has no hundreds so it is the smallest.

Now find the next smallest, and so on. ⟶ 894   Two numbers have 8 hundreds.
894 is smaller than 899.

Now write the numbers in order. ⟶ 99, 894, 899, 906, 910

WATCH OUT!
Check whether the question asks you to write them smallest first or largest first.

- To write a number to the **nearest 10,** look at the **units digit.**
  If it is 5 or more, **round up**. If it is less than 5, **round down**.

- To write a number to the **nearest 100**, look at the **tens digit.**
  If it is 5 or more, round up. If it is less than 5, round down.

- To write a number to the **nearest 1000**, look at the **hundreds digit.**
  If it is 5 or more, round up. If it is less than 5, round down.

Example

Write 37 385 to the nearest hundred.

Grade G

Look at the tens column. ———→ 37 385

37 385 to the nearest 100 is 37 400.

TIP
8 is more than 5, so round up.
The 300 becomes 400.

Practice

1. Write 75 203 in words.

   Grade G

2. Seventeen million, three hundred and fifty-four thousand people watched
   Coronation Street last Thursday.
   Write this number in figures.

   Grade G

3. Write down the value of the 8 in 56 850

   Grade G

4. Write these numbers in order of size, smallest first:

   88,   72,   302,   39,   267

   Grade G

5. Write 56 840

   (a) to the nearest thousand   **(b)** to the nearest hundred.

   Grade G

Check your answers on page 165. For full worked solutions see the CD,
See the Student Book on the CD if you need more help.

Question	1	2	3	4	5
Grade	G	G	G	G	G
Student Book pages	U2 2–3	U2 2–3	U2 1	U2 2–3	US 8–9

Negative numbers

STAGE 1

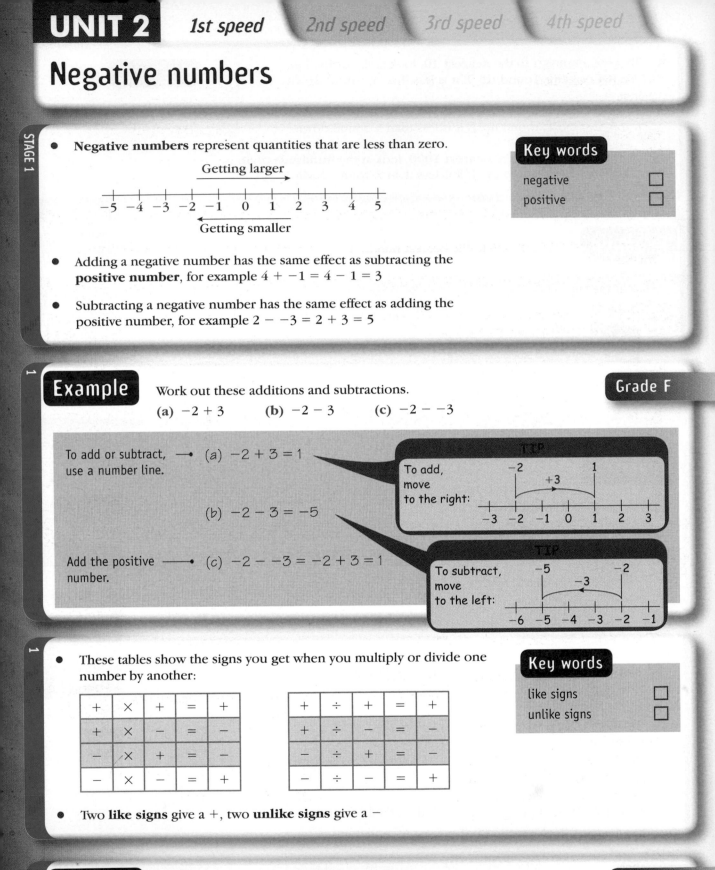

- **Negative numbers** represent quantities that are less than zero.

Getting larger

$$-5 \quad -4 \quad -3 \quad -2 \quad -1 \quad 0 \quad 1 \quad 2 \quad 3 \quad 4 \quad 5$$

Getting smaller

Key words

| negative | ☐ |
| positive | ☐ |

- Adding a negative number has the same effect as subtracting the **positive number**, for example $4 + -1 = 4 - 1 = 3$

- Subtracting a negative number has the same effect as adding the positive number, for example $2 - -3 = 2 + 3 = 5$

Example

Work out these additions and subtractions.

(a) $-2 + 3$ (b) $-2 - 3$ (c) $-2 - -3$

Grade F

To add or subtract, → (a) $-2 + 3 = 1$
use a number line.

TIP

To add,
move
to the right:

$$-3 \quad -2 \quad -1 \quad 0 \quad 1 \quad 2 \quad 3$$

(b) $-2 - 3 = -5$

Add the positive → (c) $-2 - -3 = -2 + 3 = 1$
number.

TIP

To subtract,
move
to the left:

$$-6 \quad -5 \quad -4 \quad -3 \quad -2 \quad -1$$

- These tables show the signs you get when you multiply or divide one number by another:

+	×	+	=	+
+	×	−	=	−
−	×	+	=	−
−	×	−	=	+

+	÷	+	=	+
+	÷	−	=	−
−	÷	+	=	−
−	÷	−	=	+

Key words

| like signs | ☐ |
| unlike signs | ☐ |

- Two **like signs** give a +, two **unlike signs** give a −

Example

Work out these multiplications and divisions.

(a) $-12 \div -3$ (b) $12 \div -3$ (c) 2×3

Grade E

First do the calculation ignoring the signs.

(a) $12 \div 3 = 4$ $-12 \div -3 = +4$

TIP
$- \div - = +$

Then use the 'like and unlike signs' rule to work out the sign for the answer.

(b) $12 \div 3 = 4$ $12 \div -3 = -4$

TIP
$+ \div - = -$

(c) $2 \times 3 = 6$

TIP
$+ \times + = +$

- As the temperature rises, the numbers get higher.
- As the temperature falls, the numbers get lower.

Example

Grade F

The temperature falls by 5° from the one shown on the thermometer. What is the new temperature?

10°

0°

−10

$3 - 5 = -2$

The new temperature is $-2\,°C$.

TIP
Start at the temperature shown and count down 5

+4°
+3°
+2°
+1°
0°
−1°
−2°
−3°

▼5

Practice

1 Here is a list of numbers:
   2, −10, 0, −6, −2, 6, 10

   (a) Write down the largest number.   **Grade G**

   (b) Write the numbers in order, smallest number first.   **Grade F**

2 Work out these additions and subtractions.   **Grade F**
   (a) 5 − 6   (b) −5 − 6
   (c) −5 + 6   (d) 5 + −6
   (e) 5 − −6   (f) −5 − −6

3 Last night the temperature was −6 °C.   **Grade F**
   The temperature rose by 8°.
   What was the new temperature?

4 Work out these multiplications and divisions.   **Grade E**
   (a) $+4 \times -5$ (b) $-4 \times +5$
   (c) -4×-5 (d) $20 \div -5$
   (e) $-20 \div +5$ (f) $-20 \div -5$
   (g) $\dfrac{-24}{-6}$ (h) $\dfrac{20}{-4}$

Check your answers on page 165. For full worked solutions see the CD.
See the Student Book on the CD if you need more help.

Question	1a	1b	2	3	4
Grade	G	F	F	F	E
Student Book pages		U2 12–13	U2 13–15	U2 13–15	U2 13–15

Indices and powers

STAGE 2

- The **power** is how many times a number is multiplied by itself, for example
  $2 \times 2 \times 2 \times 2 = 2^4$. You say '2 to the power 4'.

- A power is also called an **index** (plural **indices**).

- The **square** of 4 is $4 \times 4 = 4^2 = 16$
  $\sqrt{16} = 4$ means the **square root** of 16 is 4
  -4 is also a square root of 16 since $-4 \times -4 = 16$.
  There are two square roots, positive and negative.

- The **cube** of 4 is $4 \times 4 \times 4 = 4^3 = 64$
  $\sqrt[3]{64} = 4$ means the **cube root** of 64 is 4

- You can use a calculator to find squares, cubes, square roots and cube roots.

Key words

power	☐	square root	☐
index	☐	cube	☐
indices	☐	cube root	☐
square	☐		

Example

Write down the value of **(a)** 5^3 **Grade F** **(b)** $\sqrt{49}$ **Grade E**

Write out the calculation in full. ⟶ (a) $5^3 = 5 \times 5 \times 5 = 25 \times 5 = 125$

(b) $\sqrt{49} = \sqrt{7 \times 7} = 7$, $\sqrt{49} = \sqrt{-7 \times -7} = -7$

- To **multiply** powers of the same number, add the indices: $3^4 \times 3^2 = 3^{4+2} = 3^6$

- To **divide** powers of the same number, subtract the indices: $4^6 \div 4^2 = 4^{6-2} = 4^4$

Example

Simplify **(a)** $2^3 \times 2^4$ **(b)** $5^6 \div 5^4$ **Grade C**

Method 1
Write out the calculation in full. ⟶ (a) $2^3 = 2 \times 2 \times 2$
$2^4 = 2 \times 2 \times 2 \times 2$
$2^3 \times 2^4 = 2 \times 2 \times 2 \times 2 \times 2 \times 2 \times 2 = 2^7$

(b) $5^6 = 5 \times 5 \times 5 \times 5 \times 5 \times 5$
$5^4 = 5 \times 5 \times 5 \times 5$
$5^6 \div 5^4 = \dfrac{5 \times 5 \times \cancel{5} \times \cancel{5} \times \cancel{5} \times \cancel{5}}{\cancel{5} \times \cancel{5} \times \cancel{5} \times \cancel{5}} = 5^2$

Method 2
Use the rule. ⟶ (a) $2^3 \times 2^4 = 2^{3+4} = 2^7$

(b) $5^6 \div 5^4 = 5^{6-4} = 5^2$

- **BIDMAS** is a made-up word to help you remember the order of operations:

$$B I D M A S$$

Brackets   Indices   Divide   Multiply   Add   Subtract

When the operations are the same, you do them in the order they appear.

Example

Work out   **(a)** 3×5^2 **(b)** $5(7 - 4)$ **Grade E**   **(c)** $(4^2)^3$ **Grade C**

Work out the Index, then Multiply.   ⟶   (a) $3 \times 5^2 = 3 \times 25 = 75$

> **TIP**
> Remember BIDMAS!

Work out the Brackets first, then Multiply.   ⟶   (b) $5(7 - 4) = 5 \times 3 = 15$

> **TIP**
> $5(7 - 4)$ means 'multiply the contents of the bracket by 5'.

Work out the brackets first, then the index.   ⟶   (c) $(4^2)^3 = (4 \times 4)^3$
$$= 16^3 = 16 \times 16 \times 16 = 4096$$

> **TIP**
> Use long multiplication twice. For more on long multiplication see pages 44–45.

Practice

1   Write down the value of   **(a)** 3^3 **Grade F**   **(b)** $\sqrt{36}$ **Grade E**

2   Work out   **(a)** 4×5^3 **(b)** $3(9 - 4)$ **Grade E**

3   Add brackets to this calculation to make it into a true statement: **Grade D**

     $3 \times 4 + 5 = 3^3$

4   Simplify   **(a)** $4^3 \times 4^2$ **(b)** $9^7 \div 9^4$ **(c)** 7^0 **(d)** $(4^2)^3$ **Grade C**

Check your answers on page 165. For full worked solutions see the CD.
See the Student Book on the CD if you need more help.

Question	1a	1b	2	3	4
Grade	F	E	E	D	C
Student Book pages		U2 47	U2 47–49	U2 49–50	U2 51–52

Multiples, factors and primes

STAGE 1

- The **factors** of a number are whole numbers that divide exactly into the number. The factors include 1 and the number itself.

- **Multiples** of a number are the results of multiplying the number by a positive whole number.

- A **prime number** is a whole number greater than 1 which has only two factors: itself and 1.
  1 is not a prime number as it can only be divided by one number (itself).

- A **prime factor** is a factor that is a prime number.

Key words

prime number	☐
prime factor	☐
multiple	☐
factor	☐
product of prime factors	☐

Example   Here is a list of numbers: 1, 9, 12, 6, 2, 3, 24, 5

Grade F/E

Write down the numbers that are

(a) prime numbers   (b) factors of 12   (c) multiples of 6   (d) prime factors of 6

Look for numbers whose only ———• (a) 2, 3 and 5
factors are 1 and the number itself.

> **WATCH OUT!**
> Write *all* the factors — don't forget the number itself (if it is in the list). Students often forget that any number is a factor of itself and they often leave out 1.

Look for numbers that divide ———• (b) 1, 2, 3, 6 and 12
exactly into 12.

Look for numbers in the 6 times table.—• (c) 6, 12 and 24

Use your list of prime numbers ———• (d) 2 and 3
from part (a).
Which ones are factors of 6?

Example   Write each of these numbers as a **product of its prime factors**.

Grade C

(a) 12   (b) 18

Method 1
Find the prime factors by ——•
using factor trees.

▼

(a)   12

2   6

> **WATCH OUT!**
> Don't forget this step. Product means 'numbers multiplied together'.

2   3

(b)   18

2   9

3   3

Write the number as a ——•
product of its prime factors.

$12 = 2 \times 2 \times 3$
$= 2^2 \times 3$

$18 = 2 \times 3 \times 3$
$= 2 \times 3^2$

Method 2

Find the prime factors by dividing ——→
by each prime number in turn.

(a) $12 \div 2 = 6$ $\qquad$ $6 \div 2 = 3$ $\qquad$ **3** *is prime*
$\quad 12 = 2 \times 2 \times 3 = 2^2 \times 3$

(b) $18 \div 2 = 9$ $\qquad$ $9 \div 3 = 3$ $\qquad$ **3** *is prime*
$\quad 18 = 2 \times 3 \times 3 = 2 \times 3^2$

- The **highest common factor** (HCF) of two numbers is the highest factor that is common to both of them.

- The **lowest common multiple** (LCM) of two numbers is the lowest multiple that is common to both of them (or the lowest number that is a multiple of them both).

Key words

| lowest common multiple (LCM) | ☐ |
| highest common factor (HCF) | ☐ |

Example

(a) Find the highest common factor (HCF) of 12 and 18.

(b) Find the lowest common multiple (LCM) of 12 and 18.

Grade C

Write each number as a product of its prime factors.
Circle the common factors and multiply them together.

(a) $12 = ②\times 2 \times ③$
$\quad 18 = ②\times 3 \times ③$
The HCF is **2** × **3** = 6

List the first few multiples of each number.
Circle the lowest number that appears in both lists.

(b) $\quad 12 \qquad 24 \qquad ㊱ \qquad 48 \qquad 60 \qquad 72$
$\quad 18 \qquad ㊱ \qquad 54 \qquad 72$
The LCM is 36

Practice

1 Here is a list of numbers:

$\quad$ 8, 16, 5, 4, 3, 18

Write down the numbers that are

(a) factors of 16 $\qquad$ **Grade F**

(b) multiples of 4 $\qquad$ **Grade F**

(c) prime numbers. $\qquad$ **Grade E**

2 Write each of these numbers as a product of its prime factors.

(a) 16 $\qquad$ (b) 24

Grade C

3 Find the HCF of 16 and 24 $\qquad$ **Grade C**

4 Find the LCM of 16 and 24 $\qquad$ **Grade C**

Check your answers on page 165. For full worked solutions see the CD.
See the Student Book on the CD if you need more help.

Question	1ab	1c	2	3	4
Grade	F	E	C	C	C
Student Book pages		U2 9–10	U2 10–11	U2 10–11	U2 10–11

Calculating and estimating

STAGE 1

- **Sum**, **plus**, **total** and **add** are all words that mean **addition** (+).

- **Minus**, **take away** and **difference** are all words that mean **subtraction** (−).

- **Product** and **times** are words that mean **multiplication** (×).

- **Sharing** and **goes into** are words that mean **division** (÷).

Key words

addition	☐
subtraction	☐
multiplication	☐
division	☐

1

Example   Work out 324×56

Grade E

Traditional method

Multiply 324 by 6.

▼

Multiply 324 by 50. Write a zero in the units column and multiply by 5.

▼

Then add.

$$
\begin{array}{r}
3\ 2\ 4 \\
5\ 6\ \times \\
\hline
1\ 9\ 4\ 4 \\
1\ 6\ 2\ 0\ 0\ + \\
\hline
1\ 8\ 1\ 4\ 4 \\
\end{array}
$$

Grid method

×	300	20	4	×
	15 000	1000	200	50
	18 00	120	24	6

↓   ↓   ↓

16 800   1120   224 = 18 144

1

Example   Work out $544 \div 16$

Grade E

Traditional method

16 divides into 54, **3** times remainder 6.

▼

16 divides into 64, 4 times.

$$
\begin{array}{r}
3\ 4 \\
16\overline{)5\ 4\ 4} \\
4\ 8\ \downarrow \\
\hline
6\ 4 \\
6\ 4 \\
\end{array}
$$

EXAMINER'S TIP

You will find questions like this on the non-calculator paper. Make sure you get plenty of practice.

Chunking method

Subtract 10 lots of 16 (that is 10 × 16 = 160). Repeat as many times as you can.

▼

Subtract 16. Repeat as many times as you can.

▼

Add up the parts of your answer.

$$
\begin{array}{rr}
5\ 4\ 4 & \\
1\ 6\ 0\ - & 1\ 0 \\
3\ 8\ 4 & \\
1\ 6\ 0\ - & 1\ 0 \\
2\ 2\ 4 & \\
1\ 6\ 0\ - & 1\ 0 \\
6\ 4 & \\
1\ 6\ - & 1 \\
4\ 8 & \\
1\ 6\ - & 1 \\
3\ 2 & \\
1\ 6\ - & 1 \\
1\ 6 & \\
1\ 6\ - & 1 \\
0 & \\
\end{array}
$$

(3 on first bracket, 4 on second bracket)

$3 \times 10 + 4 \times 1$
$= 30 + 4 = 34$

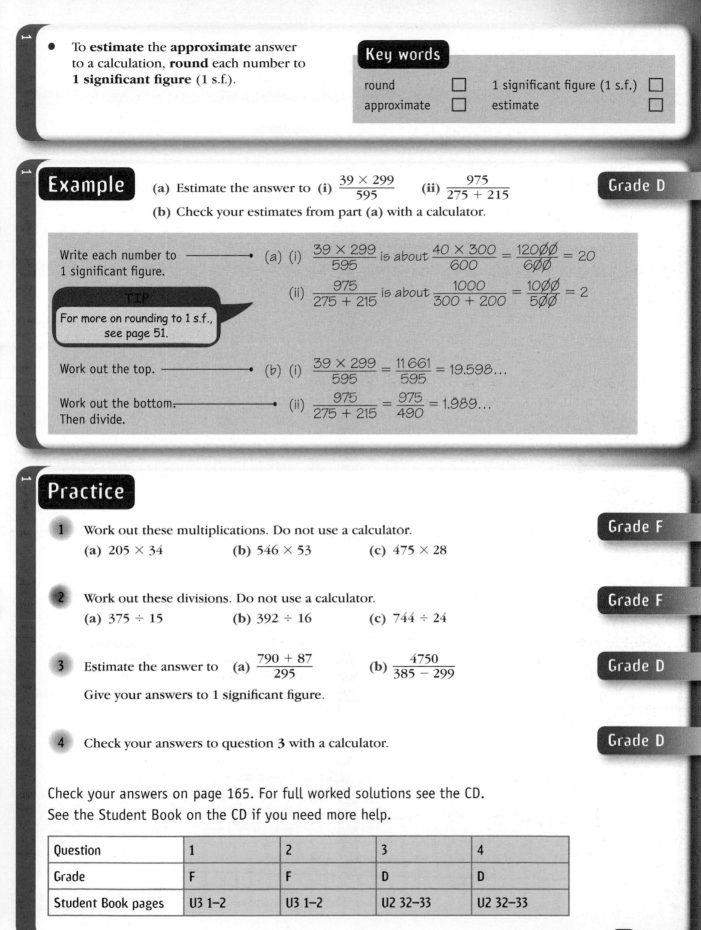

- To **estimate** the **approximate** answer to a calculation, **round** each number to **1 significant figure** (1 s.f.).

Key words

| round ☐ | 1 significant figure (1 s.f.) ☐ |
| approximate ☐ | estimate ☐ |

Example

(a) Estimate the answer to (i) $\dfrac{39 \times 299}{595}$ (ii) $\dfrac{975}{275 + 215}$

(b) Check your estimates from part (a) with a calculator.

Grade D

Write each number to 1 significant figure.

(a) (i) $\dfrac{39 \times 299}{595}$ is about $\dfrac{40 \times 300}{600} = \dfrac{1200\cancel{0}}{60\cancel{0}} = 20$

TIP

For more on rounding to 1 s.f., see page 51.

(ii) $\dfrac{975}{275 + 215}$ is about $\dfrac{1000}{300 + 200} = \dfrac{100\cancel{0}}{50\cancel{0}} = 2$

Work out the top.

(b) (i) $\dfrac{39 \times 299}{595} = \dfrac{11\,661}{595} = 19.598\ldots$

Work out the bottom. Then divide.

(ii) $\dfrac{975}{275 + 215} = \dfrac{975}{490} = 1.989\ldots$

Practice

1 Work out these multiplications. Do not use a calculator.

(a) 205×34 (b) 546×53 (c) 475×28

Grade F

2 Work out these divisions. Do not use a calculator.

(a) $375 \div 15$ (b) $392 \div 16$ (c) $744 \div 24$

Grade F

3 Estimate the answer to (a) $\dfrac{790 + 87}{295}$ (b) $\dfrac{4750}{385 - 299}$

Give your answers to 1 significant figure.

Grade D

4 Check your answers to question **3** with a calculator.

Grade D

Check your answers on page 165. For full worked solutions see the CD.
See the Student Book on the CD if you need more help.

Question	1	2	3	4
Grade	F	F	D	D
Student Book pages	U3 1–2	U3 1–2	U2 32–33	U2 32–33

Integers: topic test

Check how well you know this topic by answering these questions.
First cover the answers on the facing page.

Test questions

STAGE 1

1 (a) Write 860 245 in words.
(b) Write 3 million, two hundred and forty thousand, five hundred and six in figures.

2 Put these numbers in order, starting with the smallest number:
67, 76, 8, 88, 79

3 Write 356 276
(a) to the nearest thousand
(b) to the nearest hundred.

4 Write down the value of the 5 in 25 360

5 Find the number that is
(a) 6 less than 8
(b) 3 less than 0
(c) 9 bigger than −5
(d) 10 greater than −15

6 Find the temperature when
(a) 3 °C falls by 6°   (b) −4 °C falls by 2°
(c) −1 °C rises by 7°   (d) −13 °C rises by 5°

7 The temperature one day at noon was −1 °C. The temperature fell 7° by midnight. What was the temperature at midnight?

8 Work out these additions and subtractions.
(a) +9 + −17   (b) −3 + −1
(c) −12 + +3   (d) −6 + −6
(e) −4 − −5   (f) −11 − −4

9 Write down the value of
(a) 5^3 (b) $\sqrt{81}$

10 Work out these multiplications and divisions.
(a) +8 ÷ +2   (b) −3 × +4
(c) +10 ÷ −5   (d) −8 × −8
(e) −24 ÷ −6   (f) +6 × −9

11 James puts bottles into boxes. Each box contains 24 bottles. One day James filled 255 boxes.
(a) How many bottles did James put into boxes?
(b) The next day James put 984 bottles into boxes. How many boxes did James fill with bottles?

12 1690 football fans booked to go by coach to see their team play away. Each coach holds 57 people. How many coaches will be needed to take all the fans?

13 Here is a list of numbers:
12, 18, 6, 7, 3, 9
Write down the numbers that are
(a) factors of 18   (b) multiples of 6
(c) prime numbers   (d) prime factors of 18

STAGE 2

14 Work out
(a) 3×6^2 (b) $6(5 + 4)$

15 (a) Work out the approximate answer for
(i) $\dfrac{399 \times 52}{495}$ (ii) $\dfrac{589 + 310}{380 - 96}$
(b) Check your estimates from part (a) with a calculator.

16 Simplify
(a) $5^4 \times 5^5$ (b) $3^8 \div 3^5$ (c) $\dfrac{4^7}{4^3}$

17 Write each of these numbers as a product of its prime factors.
(a) 18   (b) 24

18 Find
(a) the LCM and
(b) the HCF of 18 and 24

Now check your answers – see the facing page.

Cover this page while you answer the test questions opposite.

Worked answers

Revise this on...

1 (a) Eight hundred and sixty thousand, two hundred and forty-five — page 36 G
(b) 3 240 506

2 8, 67, 76, 79, 88 — page 36 G

3 (a) 356 000   (b) 356 300 — page 37 G

4 5 thousands or 5000 — page 36 G

5 (a) $8 - 6 = 2$ (b) $0 - 3 = -3$ (c) $-5 + 9 = 4$ (d) $-15 + 10 = -5$ page 38 G/F

6 (a) $3 - 6 = -3\,°C$ (b) $-4 - 2 = -6\,°C$ — page 39 F
(c) $-1 + 7 = 6\,°C$ (d) $-13 + 5 = -8\,°C$

7 $-1 - 7 = -8\,°C$ — page 39 F

8 (a) -8 (b) -4 (c) -9 (d) -12 (e) 1 (f) -7 page 38 F

9 (a) 125   (b) 9 or -9 — page 40 F

10 (a) 4 (b) -12 (c) -2 (d) 64 (e) 4 (f) -54 page 38 E

11 (a) $255 \times 24 = 6120$ bottles   (b) $984 \div 24 = 41$ boxes — page 44 E

12 $1690 \div 57 = 29$ remainder 37, so 30 coaches will be needed. — page 44 E

13 (a) 3, 6, 9 and 18   (b) 6, 12 and 18   (c) 3 and 7   (d) 3 page 42 F/E/D

14 (a) $3 \times 36 = 108$ (b) $6 \times 9 = 54$ — page 41 D

15 (a) (i) $\dfrac{400 \times 50}{500} = 40$ (ii) $\dfrac{600 + 300}{400 - 100} = \dfrac{900}{300} = 3$ page 45 D

(b) (i) $20\,748 \div 495 = 41.915...$ (ii) $899 \div 284 = 3.165...$

16 (a) $5^{4+5} = 5^9$ (b) $3^{8-5} = 3^3$ (c) $4^{7-3} = 4^4$ page 40 C

17 (a) $18 = 2 \times 3 \times 3$ (b) $24 = 2 \times 2 \times 2 \times 3$ — page 42 C

18 (a) Multiples of 18: 18, 36, 54, 72, 90   Multiples of 24: 24, 48, 72, 96 page 43 C
The LCM is 72
(b) The HCF is $2 \times 3 = 6$

Tick the questions you got right.

Question	1	2	3	4	5a	5bcd	6	7	8	9	10	11	12	13ab	13c	13d	14	15	16	17	18
Grade	G	G	G	G	G	F	F	F	F	F	E	E	E	F	E	D	D	D	C	C	C

Mark the grade you are working at on your revision planner on page ix.

Adding, subtracting, multiplying and dividing decimals

STAGE 1

- When working out a decimal **addition** or **subtraction**, write the numbers in columns so that the **decimal points** are underneath one another.

- The decimal point in the answer will be underneath the ones in the calculation.

Key words

add ☐
subtract ☐
decimal point ☐

Example

Work out **(a)** $2.2 + 3.09 + 15$ **(b)** $3.2 - 1.86$

Grade G

Put the numbers in columns with the decimal points underneath one another. ———→

```
(a)    2 . 2
       3 . 0 9
      1 5 .     +
      2 0 . 2 9
           1
```

TIP

For a whole number, the decimal point goes after the units digit.

Put the numbers in columns with the decimal points underneath one another. ———→

```
(b)    3 . 2
       1 . 8 6 −
```

Put a zero to fill in the empty space. ———→
Take away as usual.

```
   3 . 2 0          ²3̷ . ¹1̷²1̷0
   1 . 8 6 −          1 . 8 6 −
                      1 . 3 4
```

EXAMINER'S TIP

Don't try to work it out in your head. Put the numbers in columns and line up the decimal points.

- When **multiplying** decimals, the answer must have the same number of **decimal places** as the total number of decimal places in the numbers being multiplied.

- Work out the multiplication without the decimal points, then put the decimal point in the answer.

Key words

multiply ☐
decimal place ☐

Example

Work out 5.26×3.4

Grade C

▶ Use the method from page 10 to work out the multiplication without decimal points.

▶ Count the total number of decimal places in the numbers you are multiplying.

▶ Put the decimal point in the answer so it has this number of decimal places.

```
      5 2 6
        3 4 ×
    2 1 0 4
  1 5 7 8 0
  1 7 8 8 4
```

5.26×3.4
2 d.p. + 1 d.p. = 3 d.p.

The answer must have 3 d.p. so it is 17.884

- When **dividing** decimals by decimals make sure you always divide by a whole number. You do this by multiplying both numbers in the division by 10 or 100 or 1000 etc.

Key word

divide ☐

Example   Work out **(a)** $12 \div 0.4$ **(b)** $3.2 \div 0.25$

Grade C

Multiply both numbers by the same number so that you are dividing by a whole number.

(a) $12 \div 0.4$
$= 120 \div 4$
$= 30$

TIP
You need to make 0.4 into 4.
Multiply *both* numbers by 10.

(b) $3.2 \div 0.25$
$= 320 \div 25$
$= 12.8$

TIP
You need to make 0.25 into 25.
Multiply *both* numbers by 100.

TIP
For more on long division see pages 44–45.

TIP
Use long division:

```
        12.8
 25)320.0
     25
     70
     50
    200
    200
```

Practice

1 Work out these additions. Show all your working.

Grade G

(a) $1.2 + 0.45 + 14$ (b) $5.07 + 0.98 + 10$ (c) £3.46 + £5.30 + £12 + £2.09

2 Work out these subtractions. Show all your working.

Grade G

(a) $4.56 - 3.8$ (b) $12.4 - 3.25$ (c) £10 − £7.24

3 Work out these multiplications. Show all your working.

(a) 5.4×7 Grade E   (b) 24.5×2.7 Grade D   (c) 3.46×0.62 Grade D

4 Work out these divisions. Show all your working.

(a) $4.5 \div 0.5$ (b) $1.28 \div 0.4$ (c) $2.88 \div 0.24$

Check your answers on page 165. For full worked solutions see the CD.
See the Student Book on the CD if you need more help.

Question	1	2	3a	3bc	4
Grade	G	G	E	D	D
Student Book pages	U2 24–25	U2 24–24	U2 27–29		U2 27–29

Rounding decimals

STAGE 2

- To **round** to a given number of **decimal places (d.p.)**, count the number of decimal places from the decimal point.

- Look at the next digit after the one you want. If it is 5 or more, **round up**. If it is less than 5, **round down**.

Key words

rounding ☐   decimal places ☐
round up ☐   round down ☐

Example

Write these numbers correct to 2 decimal places.

(a) 4.679 23   (b) 5.234 78   (c) 2.895

Grade F

Count 2 digits from the decimal point
Look at the next digit.

(a) 4.67**9** 23
   = 4.68 (to 2 d.p.)

> **TIP**
> 9 is more than 5, so round up.
> The 7 becomes 8

(b) 5.234 78
   = 5.23 (to 2 d.p.)

> **TIP**
> The next digit is 4 so round down.
> The 3 stays the same.

(c) 2.89**5**
   = 2.90 (to 2 d.p.)

> **TIP**
> The next digit is 5, so round up. 0.89 rounds up to 0.90. Keep the zero because you need 2 d.p.

- Write answers to money calculations to the nearest penny (2 d.p.).

Example

David bought 10 litres of fuel at three different garages.
   At Garage A it cost £18
   At Garage B it cost £2.10 more than at Garage A.
   At Garage C it cost £20.19

Work out the cost of 1 litre of fuel at each garage.

Grade F

Divide the total cost by the number of litres.
Write the answer to the nearest penny (2 d.p.).

Garage A: £18 ÷ 10 = £1.8
         1 litre costs £1.80

Garage B: Cost for 10 litres
         = £18 + £2.10 = £20.10
         £20.10 ÷ 10 = £2.01
         1 litre costs £2.01

Garage C: £20.19 ÷ 10 = £2.019
         1 litre costs £2.02 (to 2 d.p.)

> **WATCH OUT!**
> You need to add a zero so there are 2 d.p.
> £1.80 means 1 pound and 80 pence.
> £1.08 means 1 pound and 8 pence.
> The position of the zero is important!

> **EXAMINER'S TIP**
> If a question doesn't tell you how to give a decimal answer, give it to either 3 significant figures or 2 decimal places or the nearest penny.

- To round to a given number of **significant figures** (s.f.), count the number of digits from the first non-zero digit, starting from the *left*.

- Look at the next digit after the one you want.
  If it is **5** or more, round up. If it is less than **5**, round down.

- Use zeros to show the **place value**.

Example Write these numbers correct to 3 significant figures.

(a) 647 485   (b) 1 765 891   (c) 0.004 675 26

Grade E

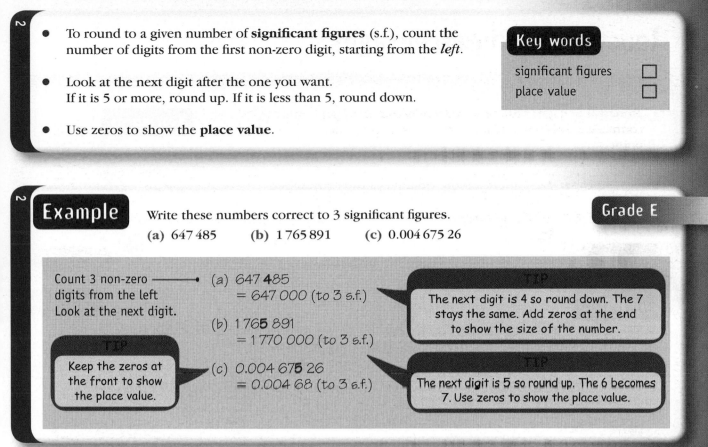

Count 3 non-zero digits from the left. Look at the next digit.

(a) 647 **4**85
= 647 000 (to 3 s.f.)

TIP

The next digit is **4** so round down. The 7 stays the same. Add zeros at the end to show the size of the number.

(b) 1 76**5** 891
= 1 770 000 (to 3 s.f.)

TIP

Keep the zeros at the front to show the place value.

(c) 0.004 67**5** 26
= 0.004 68 (to 3 s.f.)

TIP

The next digit is **5** so round up. The 6 becomes 7. Use zeros to show the place value.

Practice

1 Rana has ten pounds fifty pence and Axel has eight pounds six pence.
Write these amounts in figures.

Grade G

2 Terry buys 20 litres of petrol for £18.
Work out the cost of 1 litre of petrol.

Grade F

3 Write these numbers correct to
(a) 1 decimal place   (b) 2 decimal places.
(i) 5.4523   (ii) 10.398   (iii) 4.0562

Grade F

4 Write these numbers correct to
(a) 1 significant figure   (b) 3 significant figures.
(i) 250 398   (ii) 56 921   (iii) 0.347 23   (iv) 0.000 599 772

Grade E

Check your answers on page 165. For full worked solutions see the CD.
See the Student Book on the CD if you need more help.

Question	1	2	3	4
Grade	G	F	F	E
Student Book pages	U2 22–23	U2 27–29	U2 30–31	U2 30–31

Fractions and percentages

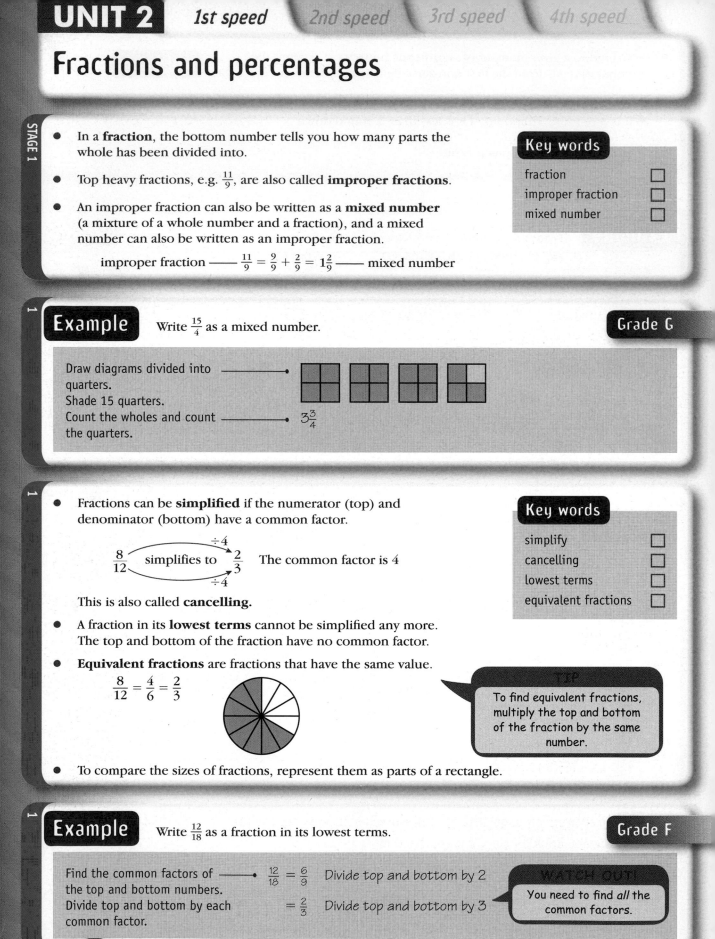

STAGE 1

- In a **fraction**, the bottom number tells you how many parts the whole has been divided into.

- Top heavy fractions, e.g. $\frac{11}{9}$, are also called **improper fractions**.

- An improper fraction can also be written as a **mixed number** (a mixture of a whole number and a fraction), and a mixed number can also be written as an improper fraction.

improper fraction —— $\frac{11}{9} = \frac{9}{9} + \frac{2}{9} = 1\frac{2}{9}$ —— mixed number

Key words
- fraction ☐
- improper fraction ☐
- mixed number ☐

Example   Write $\frac{15}{4}$ as a mixed number.   Grade G

Draw diagrams divided into quarters.
Shade 15 quarters.
Count the wholes and count the quarters.

$3\frac{3}{4}$

- Fractions can be **simplified** if the numerator (top) and denominator (bottom) have a common factor.

$\frac{8}{12}$ simplifies to $\frac{2}{3}$ The common factor is 4

$\div 4$

This is also called **cancelling**.

- A fraction in its **lowest terms** cannot be simplified any more. The top and bottom of the fraction have no common factor.

- **Equivalent fractions** are fractions that have the same value.

$\frac{8}{12} = \frac{4}{6} = \frac{2}{3}$

Key words
- simplify ☐
- cancelling ☐
- lowest terms ☐
- equivalent fractions ☐

TIP
To find equivalent fractions, multiply the top and bottom of the fraction by the same number.

- To compare the sizes of fractions, represent them as parts of a rectangle.

Example   Write $\frac{12}{18}$ as a fraction in its lowest terms.   Grade F

Find the common factors of the top and bottom numbers.
Divide top and bottom by each common factor.

$\frac{12}{18} = \frac{6}{9}$ Divide top and bottom by 2

$= \frac{2}{3}$ Divide top and bottom by 3

WATCH OUT!
You need to find *all* the common factors.

52

Example

Write two fractions that are equivalent to $\frac{3}{4}$

Multiply top and bottom of
the fraction by the same
number.

$\frac{3}{4} = \frac{6}{8}$ Multiply top and bottom by 2

$\frac{3}{4} = \frac{9}{12}$ Multiply top and bottom by 3

Example

Which fraction is larger, $\frac{2}{3}$ or $\frac{3}{4}$?

Split the rectangle into
thirds and into quarters.
Shade 2 thirds and 3 quarters.

TIP

Split a rectangle in two
directions so you can
compare the fractions.

TIP

The lowest common multiple of
3 and 4 is 12 so your rectangle
should be split into 12 squares.

$\frac{2}{3} = 8$ parts $\frac{3}{4} = 9$ parts $\frac{3}{4}$ is larger.

Example

Work out 15% of £60

TIP

'Percent' means 'out of 100'. So 15% means 15 out of 100.

Method 1
Write the percentage as a fraction
with denominator 100 and multiply.

$\frac{15}{100} \times £60 = \frac{15 \times £60}{100} = \frac{£900}{100} = £9$

TIP

'of' means 'times'

Method 2
Divide by 10 to find 10%

10% of £60 = £60 ÷ 10 = £6

Find 5% by dividing 10% by 2

5% of £60 = £6 ÷ 2 = £3

TIP

5% is half of 10%

Add 10% and 5%

£6 + £3 = £9

Practice

1 Shade $\frac{4}{5}$ of this shape.

2 Write   Grade G

(a) $\frac{12}{5}$ as a mixed number

(b) $3\frac{2}{3}$ as an improper fraction.

3 Write $\frac{18}{24}$ as a fraction in its
lowest terms.

4 Write two fractions that are
equivalent to $\frac{5}{6}$.

5 Which fraction is larger, $\frac{4}{5}$ or $\frac{3}{4}$?

6 Work out 15% of
(a) 40 kg   (b) £120

7 Work out 40% of £80

Check your answers on page 165. For full worked solutions see the CD.
See the Student Book on the CD if you need more help.

Question	1	2	3	4	5	6	7
Grade	G	G	F	F	F	E	E
Student Book pages	U2 19–20	U2 19–20	U2 20–21	U2 20–21	U2 21–22	U2 39–40	U2 39–40

Decimals, fractions and percentages: topic test

Check how well you know this topic by answering these questions.
First cover the answers on the facing page.

Test questions

STAGE 1

1 Work out these calculations. Show all your working.
 (a) 23.5 + 12.54 + 0.66
 (b) 5.44 + 15 + 19.85
 (c) 4.87 − 1.53
 (d) 3.00 − 1.37

2 Henri saves what is left of his pocket money each week.
In the last month he saved the following amounts each week.
 £2.55, £3.27, £1.95, £3.27
How much did Henri save altogether last month?

3 Rosie bought a clock for £17.65.
She paid with a £20 note.
How much change should she get?

4 Katrin has one pound fourteen pence, Joe has one pound four pence, and Susan has one pound forty pence.
Write these amounts in figures.

5 Shade $\frac{5}{6}$ of this diagram.

6 Write $\frac{11}{4}$ as a mixed number.

7 Express these fractions in their lowest terms.
 (a) $\frac{24}{26}$ (b) $\frac{15}{20}$

8 Copy and complete these sets of equivalent fractions.
 (a) $\frac{5}{8} = \frac{}{16} = \frac{}{24} = \frac{}{32} = \frac{}{40} = \frac{}{48} = \frac{}{64}$
 (b) $\frac{2}{7} = \frac{}{14} = \frac{}{21} = \frac{}{28} = \frac{}{35} = \frac{}{63} = \frac{}{84}$

STAGE 2

9 Work out 10% of
 (a) £50   (b) 80 metres
 (c) 300 km

10 Write these numbers correct to 2 decimal places.
 (a) 5.775   (b) 0.7757
 (c) 3.999   (d) 0.779
 (e) 23.0055   (f) 25.794

11 Maria sells 45 packets of coloured gel pens for £5.49 each.
How much money does she collect altogether?

12 Josie bought 16 identical teddy bears for £68.
 (a) How much did she pay for each one?
 (b) She sold all the bears for £5.95 each. What is her total profit?

13 Keri buys 5 kg of potatoes. The total cost is £2.00
Work out the cost of 1 kg.

14 Tom buys a pack of 10 pens. The pack of pens cost £3.75
Find the cost of 1 pen.

15 Write these numbers correct to 3 significant figures.
 (a) 34 565   (b) 6885
 (c) 34.87   (d) 5.5683
 (e) 0.023 57   (f) 0.004 899

Now check your answers – see the facing page.

Cover this page while you answer the test questions opposite.

Worked answers

Revise this on...

1 (a)
```
  23.5
  12.54
   0.66+
  36.70
   1  1
```
(b)
```
   5.44
  15
  19.85+
  40.29
   2 1
```
(c)
```
  4.87
  1.53−
  3.34
```
(d)
```
  2З.⁹Ø¹0
  1. 37−
  1. 63
```
page 48   G

2
```
 £2.55
 £3.27
 £1.95
 £3.27+
£11.04
  2  2
```
3
```
£¹²Ø.⁹Ø¹0
£ 17. 65−
£  2. 35
```
page 48   G

4 Katrin £1.14, Joe £1.04, Susan £1.40   page 50   G

5

6 $\frac{11}{4} = 2\frac{3}{4}$ page 52   G

7 (a) $\frac{24}{36} = \frac{12}{18} = \frac{6}{9} = \frac{2}{3}$ (b) $\frac{15}{20} = \frac{3}{4}$ page 52   F

8 (a) $\frac{5}{8} = \frac{10}{16} = \frac{15}{24} = \frac{20}{32} = \frac{25}{40} = \frac{30}{48} = \frac{40}{64}$ page 52   F

(b) $\frac{2}{7} = \frac{4}{14} = \frac{6}{21} = \frac{8}{28} = \frac{10}{35} = \frac{18}{63} = \frac{24}{84}$

9 (a) £50 ÷ 10 = £5   (b) 80 m ÷ 10 = 8 m   (c) 300 km ÷ 10 = 30 km   page 53   F

10 (a) 5.78   (b) 0.78   (c) 4.00   page 50   E
(d) 0.78   (e) 23.01   (f) 25.79

11
```
    549
     45×
   2745
  21960
  24705
    1  1
```
£5.48 × 45 = £247.05

12 (a)
```
        4.25
  16 )68.00
        64
        40
        32
        80    £4.25 each
```
(b) £5.95 × 16 = £95.20
£95.20 − £68 = £27.20

page 49   D

13 200 ÷ 5 = 40p or £0.40

14 375 ÷ 10 = 37.5 = 38p or £0.38   page 49   D

15 (a) 34 600   (b) 6890   (c) 34.9   page 51   C
(d) 5.57   (e) 0.0236   (f) 0.004 90

Tick the questions you got right.

Question	1	2	3	4	5	6	7	8	9	10	11	12	13	14	15
Grade	G	G	G	G	G	G	F	F	F	E	D	D	D	D	C

Mark the grade you are working at on your revision planner on page ix.

Number: subject test

Check how well you know this topic by answering these questions.

Exam practice questions

STAGE 1

1 **(a)** Write 7432 in words.

(b) Round 23 250 to the nearest thousand.

(c) Write down the value of the 3 in 63 750

(d) Write twenty-four thousand, five hundred and seventy-six in figures.

2 Write $\frac{15}{20}$ as a fraction in its simplest form.

3 Arrange these numbers in order.
Start with the smallest number.

(a) 36   17   84   101   42

(b) 0.09   0.6   0.69   0.006   0.06

(c) 0   7   −3   −7   3   −2

4 Work out these.

(a) $2 + 6$

(b) $-3 - 7$

(c) -3×-2

5 Work out
(a) $3 + 4 \times 2$ **(b)** $4 \times (7 - 2)$

6 Work out these. Show all your working.

(a) 246×43

(b) $3.75 \div 0.15$

7 **(a)** Find the LCM of 24 and 36

(b) Find the HCF of 24 and 36

8 Shade in $\frac{3}{4}$ of this rectangle.

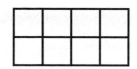

9 Work out 35% of £80

10 Find the value of

(a) 4^3

(b) $\sqrt{1.69}$

11 (a) Estimate the answer to

$$\frac{299 \times 9.78}{21 \times 0.0199}$$

(b) Use your calculator to find the answer to

$$\frac{3.45^2 - \sqrt{11.5}}{4.56 + 2.05}$$

(c) Write your answer to part (b)

(i) to 2 decimal places

(ii) to 1 significant figure.

12 Work out

(a) $7^3 \times 7^5$

(b) $5^7 \div 5^3$

Check your answers on page 165. For full worked solutions see the CD.

Tick the questions you got right.

Question	1	2	3a	3b	3c	4ab	4c	5	6a	6b	7	8	9	10	11a	11b	11ci	11cii	12
Grade	G	F	G	E	F	F	E	E	E	C	C	G	E	D	D	C	F	E	C
Revise this on page	36–37	52	36–37, 38–39	—	38	41	44	49	43	52	53	40–41	45	50	51	40–41

Mark the grade you are working at on your revision planner on page ix.

Go to the pages shown to revise for the ones you got wrong.

Simplifying algebra

STAGE 2

- An **algebraic expression** is a collection of letters, symbols and numbers:

  $a + 3b - 2c$ is an algebraic expression.

  This is a **term**

- You can combine **like terms** by adding and subtracting them:

  $2a + 3a = 5a$ and $3b + 4b - b = 6b$

- You can simplify algebraic expressions by collecting like terms together:

  $2a - 4b + 3a + 5b$ simplifies to $5a + b$

Key words

algebraic expression	☐
term	☐
like terms	☐

2

Example   Simplify $3a + 5d + 2a + 6 - 3d - 8$ **Grade E**

$$3a + 5d + 2a + 6 - 3d - 8$$

Start by collecting like terms:
all the a terms, then
the d terms, then any ———• $= 3a + 2a + 5d - 3d + 6 - 8$
separate numbers.

▼

Add or subtract ———• $=\quad 5a\qquad +2d\qquad -2$
like terms. $= 5a + 2d - 2$

TIP
Keep each sign with its
own term (no sign
means +, so 3a is +3a).
Here 5d is +5d,
and the 3d term is −3d.

TIP
$+6 + -8 = -2$
For more on adding negative
numbers see page 38.

WATCH OUT!
When collecting terms by adding or subtracting, the
letter stays the same: $a + a = 2a$.
When you multiply, $a \times a = a^2$
Students sometimes wrongly write $3a + 2a$ as $5a^2$ or $6a^2$.

2

- **Expanding** the brackets means multiplying to remove the brackets.

- **Factorising** means splitting up an expression using brackets:

  $12a + 4b = 4(3a + b)$

Key words

expand	☐
factorise	☐

2

Example   Expand $3(x + 5)$ **Grade D**

Multiply each term inside the bracket ———• $3(x + 5) = \mathbf{3} \times x + \mathbf{3} \times 5$
by the term outside the bracket. $= 3x + 15$

2

Example   Simplify $5(a + 2b) - 2(2a - 5b)$ **Grade C**

$$5(a + 2b) - 2(2a - 5b)$$

Multiply each term inside the bracket by the term outside the bracket. →
$= \mathbf{5} \times a + \mathbf{5} \times +2b - 2 \times +2a - 2 \times -5b$
$= 5a + 10b - 4a + 10b$

TIP
A minus term outside the bracket multiplying a minus term inside the bracket gives
$- \times - = +$
Here $-2 \times -5 = +10$
For more on multiplying negative numbers see pages 38–39.

Collect like terms together: → $= 5a - 4a + 10b + 10b$
all the a terms, then the b terms.

WATCH OUT!
Make sure you multiply *every* term inside the bracket by the term outside. Students often wrongly expand $5(a + 2b)$ to give $5a + 2b$ instead of the correct expansion $5a + 10b$.

Add or subtract the → $= \quad a \quad + \quad 20b$
like terms. $= a + 20b$

Example

Factorise $3x^2 - 6x$

EXAMINER'S TIP
You need to factorise *fully* to get full marks.

Grade C

Find all the common factors → $3x^2 - 6x$
of both terms. 3 and x are common factors.

Write the highest common → $= 3x(\quad)$
factor outside the bracket.

WATCH OUT!
To factorise fully, make sure you have the *highest* common factor outside the bracket. Students often wrongly use only one factor, obtaining either $3(x^2 - 2x)$ or $x(3x - 6)$.

Work out what is needed → $= 3x(x - 2)$
inside the bracket.

TIP
You can check your answer by expanding: $3x(x - 2) = 3x^2 - 6x$

Practice

1 Simplify   **Grade F**
(a) $e + e + e + e + e$ (b) $3 \times j \times k$

2 Simplify   **Grade E**
(a) $5a + 4b + 3a - 3b$
(b) $p^3 + p^3 + p^3 + p^3$

3 Simplify   **Grade D**
(a) $3p \times 5q$ (b) $t \times t \times t$

4 Expand   **Grade D**
(a) $2(c + 6)$ (b) $d(a - 3)$ (c) $b(b + 2)$

5 Expand and simplify   **Grade C**
(a) $2(5p - 3) + 4(3p + 2)$
(b) $5(3g + 2h) - 3(4g - h)$
(c) $4p + 3(p + 2)$

6 Factorise   **Grade C**
(a) $3c + 12$ (b) $4m^2 - 12m$
(c) $2t^2 - 6t$

Check your answers on page 165. For full worked solutions see the CD.
See the Student Book on the CD if you need more help.

Question	1	2	3	4	5	6
Grade	F	E	D	D	C	C
Student Book pages	U2 66–68	U2 66–67	U2 66–67, U3 59–61	U2 70	U2 70–71	U2 71–72

Number sequences

STAGE 1

- In a **number pattern** or **sequence** there is always a **rule** to get from one number to the next. For example:

1, 4, 7, 10, …	The rule is: add 3
50, 46, 42, 38, …	The rule is: take away 4
2, 4, 8, 16, …	The rule is: multiply by 2

- To find the rule for the *n*th **term** of a number pattern, use a table of values. For example:

Key words

- pattern ☐
- sequence ☐
- term ☐
- difference ☐
- rule ☐

Term number	1	2	3	4
Term	1	4	7	10
Difference		+3	+3	+3

The rule is 'add 3' so the number in front of the *n* is 3. The rule starts with $3n$.
Look at the first term, $n = 1$, to find the number to add or subtract.

1

Example

Here is a pattern made from sticks:

Diagram 1   Diagram 2   Diagram 3

(a) Draw Diagram 4.

(b) Complete the table.

Diagram	1	2	3	4	5
Number of sticks	5	9	13		

(c) Write down a formula for the number of sticks, *S*, in terms of the pattern number, *n*.

Grade G

Grade G

Grade C

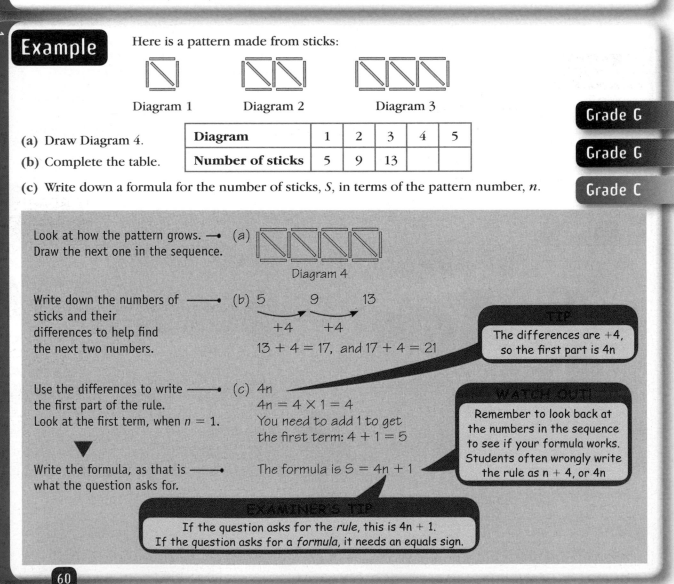

Look at how the pattern grows. → (a)

Draw the next one in the sequence.

Diagram 4

Write down the numbers of → (b) 5   9   13
sticks and their
differences to help find   +4   +4
the next two numbers. $13 + 4 = 17$, and $17 + 4 = 21$

TIP
The differences are +4, so the first part is $4n$

Use the differences to write → (c) $4n$
the first part of the rule. $4n = 4 \times 1 = 4$
Look at the first term, when $n = 1$.   You need to add 1 to get
     the first term: $4 + 1 = 5$

WATCH OUT!
Remember to look back at the numbers in the sequence to see if your formula works. Students often wrongly write the rule as $n + 4$, or $4n$

Write the formula, as that is → The formula is $S = 4n + 1$
what the question asks for.

EXAMINER'S TIP
If the question asks for the *rule*, this is $4n + 1$.
If the question asks for a *formula*, it needs an equals sign.

Example

This is part of a sequence of numbers: 2, 8, __, __, 26

(a) Find the missing numbers.

(b) Write down the general rule, in terms of n, for the nth term of the sequence.

Find the number differences. ⟶ (a) 2   8   __   __   26

+6   +6   +6   +6

Use the difference to work out ⟶ $8 + 6 = 14$, then $14 + 6 = 20$
the missing numbers.

TIP
Check that your rule gives the next term correctly:
$20 + 6 = 26$

Use the differences to write the ⟶ (b) The differences are +6, so the first part of
first part of the rule.   the rule is $6n$

Look at the first term, ⟶ $6n = 6 \times 1 = 6$
when $n = 1$.   You need to subtract 4 to get the first term: $6 - 4 = 2$

Write the rule. ⟶ $6n - 4$

Practice

1 Here is a pattern made from dots:

Diagram 1   Diagram 2   Diagram 3

(a) Draw Diagram 4.   **Grade G**

(c) Write down a rule to   **Grade F**
work out the number
of dots in the 15th
diagram.

(b) Complete the table.   **Grade G**

Diagram	1	2	3	4	5
Number of dots	7	10	13		

(d) Find a rule for the number of dots, in   **Grade C**
terms of n, in the nth diagram.

2 Here is a sequence of numbers: 30, 26, __, __, 14, 10   **Grade F**
Find the missing terms in this sequence of numbers.

3 Here is number sequence: 4, 11, 18, 25, 32   **Grade C**
Find a rule, in terms of n, for the nth term of this sequence.

Check your answers on page 166. For full worked solutions see the CD.
See the student book on the CD if you need more help.

Question	1ab	1c	1d	2	3
Grade	G	F	C	F	C
Student Book pages	U2 80–83		U2 84–87	U2 78–79	U2 84–86

Coordinates

STAGE 1

- The number line is **1-dimensional** or **1-D**. You can describe positions on the number line using one number or **coordinate**, for example (2).

- Flat shapes are **2-dimensional** or **2-D**. You can describe positions on a flat shape using two numbers or coordinates, for example (2, 1).

Key words

1-dimensional	☐	axis	☐
2-dimensional	☐	mid-point	☐
coordinates	☐		

1

Example

(a) Write down the coordinates of
   (i) point *A*   (ii) point *B*.

(b) Plot these points on the grid.
   (i)   *C*(2, 3)
   (ii) *D*(−2, 0)
   (iii) *E*(−2, −1)

(c) Write down the coordinates of the **mid-point** of the line *FG*.

Grade F

Grade F

Grade D

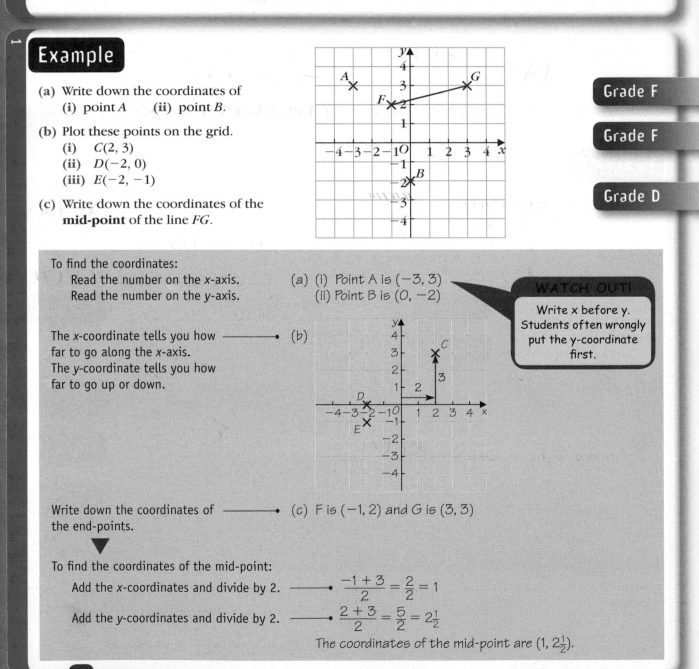

To find the coordinates:
   Read the number on the *x*-axis.
   Read the number on the *y*-axis.

(a) (i) Point A is (−3, 3)
    (ii) Point B is (0, −2)

WATCH OUT!
Write x before y. Students often wrongly put the y-coordinate first.

The *x*-coordinate tells you how ——• (b) far to go along the *x*-axis.
The *y*-coordinate tells you how far to go up or down.

Write down the coordinates of ——• (c) F is (−1, 2) and G is (3, 3)
the end-points.

▼

To find the coordinates of the mid-point:
   Add the *x*-coordinates and divide by 2. ——• $\frac{-1 + 3}{2} = \frac{2}{2} = 1$

   Add the *y*-coordinates and divide by 2. ——• $\frac{2 + 3}{2} = \frac{5}{2} = 2\frac{1}{2}$

The coordinates of the mid-point are $(1, 2\frac{1}{2})$.

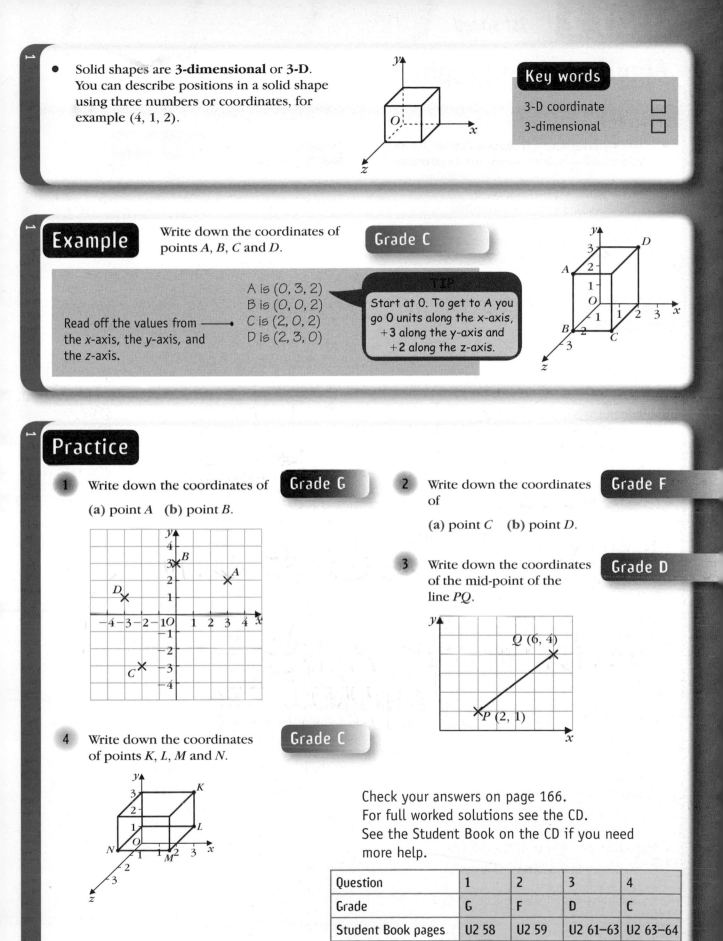

- Solid shapes are **3-dimensional** or **3-D**.
  You can describe positions in a solid shape using three numbers or coordinates, for example (4, 1, 2).

Example

Write down the coordinates of points A, B, C and D.

Grade C

Read off the values from the x-axis, the y-axis, and the z-axis.

A is $(0, 3, 2)$
B is $(0, 0, 2)$
C is $(2, 0, 2)$
D is $(2, 3, 0)$

TIP

Start at O. To get to A you go 0 units along the x-axis, +3 along the y-axis and +2 along the z-axis.

Practice

1 Write down the coordinates of

(a) point A (b) point B.

Grade G

2 Write down the coordinates of

(a) point C (b) point D.

Grade F

3 Write down the coordinates of the mid-point of the line PQ.

Grade D

Q (6, 4)

P (2, 1)

4 Write down the coordinates of points K, L, M and N.

Grade C

Check your answers on page 166.
For full worked solutions see the CD.
See the Student Book on the CD if you need more help.

Question	1	2	3	4
Grade	G	F	D	C
Student Book pages	U2 58	U2 59	U2 61–63	U2 63–64

Algebraic line graphs

STAGE 2

- The **equation of a line** uses algebra to show a relationship between the x- and y-coordinates of points on the line.

- An equation containing an x-term and no higher powers of x (such as x^2) is a **linear equation**.

- The graph of a linear equation is a straight line.

Example

Grade E

(a) Complete the table of values for $y = 3x - 1$.

x	-2	-1	0	1	2	3
y						

(b) Use your table of values to draw the graph of $y = 3x - 1$.

(c) Use your graph to find

     (i)   the value of y when $x = 1.5$

     (ii)   the value of x when $y = 6.5$

Work out each value of y.

TIP

Start working out the values from the right-hand (positive) side of the table. Try to spot the pattern in the numbers in the table: this will help you complete it. Here the pattern is -3 each time (from right to left).

(a) When $x = \mathbf{3}$, $y = 3 \times \mathbf{3} - 1 = 8$
When $x = \mathbf{2}$, $y = 3 \times \mathbf{2} - 1 = 5$
When $x = \mathbf{1}$, $y = 3 \times \mathbf{1} - 1 = 2$

x	-2	-1	0	1	2	3
y	-7	-4	-1	2	5	8

This represents the point $(-2, -7)$

Plot the table values on the grid.

▼

Join the points with a single straight line.

WATCH OUT!

$y = 3x - 1$ is a linear equation. If the points do not join to give a single straight line you have made an error: check your working.

(b)

Draw a line from the x-value up to the line and across to the y-axis.

(c) (i) When $x = 1.5$,
            $y = 3.5$

Draw a line from the y-value across to the line and down to the x-axis.

(ii) When $y = 6.5$,
          $x = 2.5$

Example

(a) Write down the equation of the line AB.

(b) On a copy of the grid, draw the line

(i) $x = 3$ (ii) $y = -2$

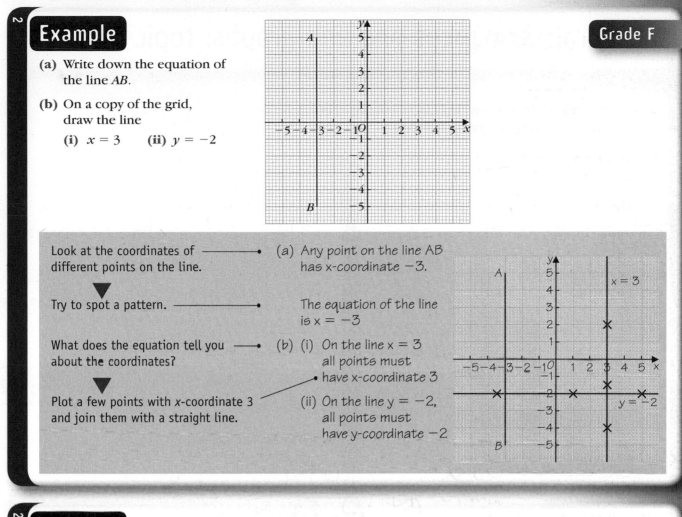

Look at the coordinates of different points on the line.

▼

Try to spot a pattern.

What does the equation tell you about the coordinates?

▼

Plot a few points with x-coordinate 3 and join them with a straight line.

(a) Any point on the line AB has x-coordinate −3.

The equation of the line is x = −3

(b) (i) On the line x = 3 all points must have x-coordinate 3

(ii) On the line y = −2, all points must have y-coordinate −2

Practice

1 On the grid, draw the line

Grade F

(a) $y = 2$ (b) $x = -4$

2 (a) Complete the table of values for $y = \frac{1}{2}x + 1$.

Grade E

x	-4	-3	-2	-1	0	1	2
y							

(b) Draw the graph of $y = \frac{1}{2}x + 1$ on the grid.

(c) Use your graph to find
(i) the value of y when $x = \frac{1}{2}$
(ii) the value of x when $y = -\frac{3}{4}$

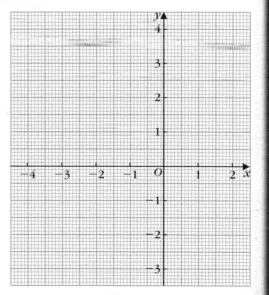

Check your answers on page 166. For full worked solutions see the CD.

See the Student Book on the CD if you need more help.

Question	1	2
Grade	F	E
Student Book pages	U2 90–91	U2 93–97

Algebra, sequences and line graphs: topic test

Check how well you know this topic by answering these questions.
First cover the answers on the facing page.

Test questions

STAGE 1

1 This is a series of diagrams made from sticks.

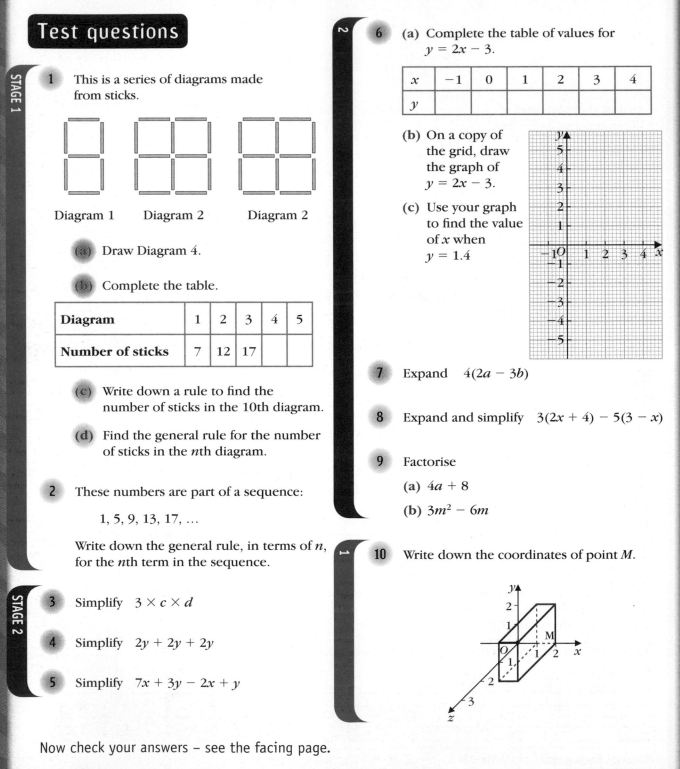

Diagram 1　　　Diagram 2　　　Diagram 2

(a) Draw Diagram 4.

(b) Complete the table.

Diagram	1	2	3	4	5
Number of sticks	7	12	17		

(c) Write down a rule to find the number of sticks in the 10th diagram.

(d) Find the general rule for the number of sticks in the nth diagram.

2 These numbers are part of a sequence:

   1, 5, 9, 13, 17, …

Write down the general rule, in terms of n, for the nth term in the sequence.

STAGE 2

3 Simplify $3 \times c \times d$

4 Simplify $2y + 2y + 2y$

5 Simplify $7x + 3y - 2x + y$

2

6 (a) Complete the table of values for $y = 2x - 3$.

x	-1	0	1	2	3	4
y						

(b) On a copy of the grid, draw the graph of $y = 2x - 3$.

(c) Use your graph to find the value of x when $y = 1.4$

7 Expand $4(2a - 3b)$

8 Expand and simplify $3(2x + 4) - 5(3 - x)$

9 Factorise

(a) $4a + 8$

(b) $3m^2 - 6m$

1

10 Write down the coordinates of point M.

Now check your answers – see the facing page.

Cover this page while you answer the test questions opposite.

Worked answers

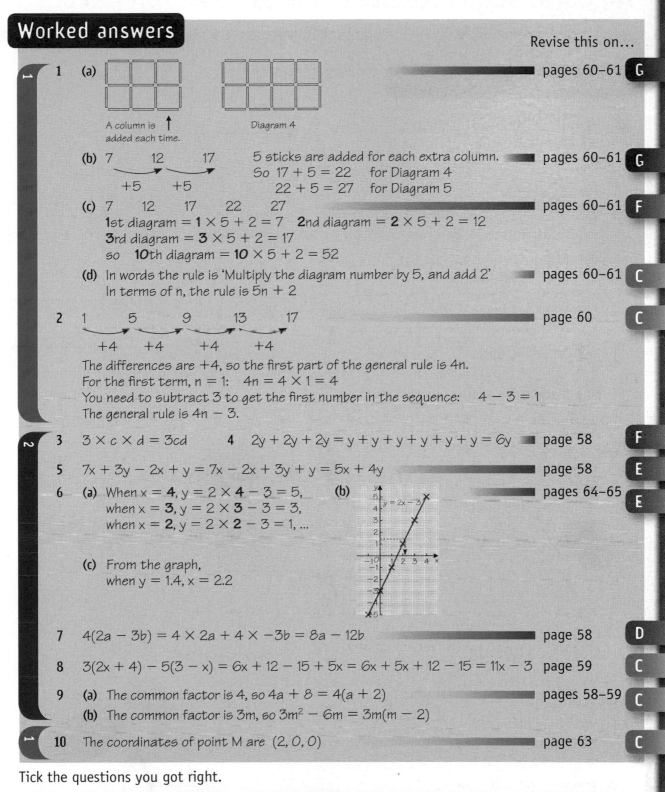

Revise this on...

1 **(a)** <!-- diagram --> pages 60–61 **G**

A column is ↑ added each time.

Diagram 4

(b) 7   12   17   5 sticks are added for each extra column.   pages 60–61 **G**
     +5   +5   So 17 + 5 = 22   for Diagram 4
                   22 + 5 = 27   for Diagram 5

(c) 7   12   17   22   27   pages 60–61 **F**
1st diagram = **1** × 5 + 2 = 7   2nd diagram = **2** × 5 + 2 = 12
3rd diagram = **3** × 5 + 2 = 17
so   **10**th diagram = **10** × 5 + 2 = 52

(d) In words the rule is 'Multiply the diagram number by 5, and add 2'   pages 60–61 **C**
In terms of n, the rule is 5n + 2

2 1   5   9   13   17   page 60 **C**
   +4   +4   +4   +4
The differences are +4, so the first part of the general rule is 4n.
For the first term, n = 1:   4n = 4 × 1 = 4
You need to subtract 3 to get the first number in the sequence: 4 − 3 = 1
The general rule is 4n − 3.

3 3 × c × d = 3cd   **4** 2y + 2y + 2y = y + y + y + y + y + y = 6y   page 58 **F**

5 7x + 3y − 2x + y = 7x − 2x + 3y + y = 5x + 4y   page 58 **E**

6 **(a)** When x = **4**, y = 2 × **4** − 3 = 5,   **(b)**   pages 64–65 **E**
      when x = **3**, y = 2 × **3** − 3 = 3,
      when x = **2**, y = 2 × **2** − 3 = 1, ...

(c) From the graph,
    when y = 1.4, x = 2.2

7 4(2a − 3b) = 4 × 2a + 4 × −3b = 8a − 12b   page 58 **D**

8 3(2x + 4) − 5(3 − x) = 6x + 12 − 15 + 5x = 6x + 5x + 12 − 15 = 11x − 3   page 59 **C**

9 **(a)** The common factor is 4, so 4a + 8 = 4(a + 2)   pages 58–59 **C**
   (b) The common factor is 3m, so 3m² − 6m = 3m(m − 2)

10 The coordinates of point M are (2, 0, 0)   page 63 **C**

Tick the questions you got right.

Question	1ab	1c	1d	2	3	4	5	6	7	8	9	10
Grade	G	F	C	C	F	F	E	E	D	C	C	C

Mark the grade you are working at on your revision planner on page x.

Algebra: subject test

STAGE 1

Exam practice questions

1 Here are some terms from a number sequence.

5, 9, 13,__, __, 25, 29

Find the terms missing from the sequence.

2 **(a)** Write down the coordinates of the points A, B, C and D.

(b) Plot these points on the grid:
E(0, −2); F(−4, 2); G(2, −3); H(−2, −3).

(c) Write down the coordinates of the midpoint of the line
(i) AB   **(ii)** CD

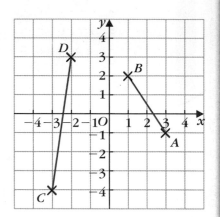

3 This is a series of diagrams made from dots.

Diagram 1   Diagram 2   Diagram 3

(a) Draw Diagram 4.

(b) Complete the table.

Diagram	1	2	3	4	5
Number of dots	4	7	10		

(c) Find the general rule, in terms of n, for the number of dots in the nth diagram.

4 These five numbers are part of a number sequence: 8, 11, 14, 17, 20

Write down the general rule, in terms of n, for the nth term in the sequence.

5 This cuboid is drawn on a 3-D grid.
A, B, C and D are vertices of the cuboid.

Write down the 3-D coordinates of the points
A, B, C and D.

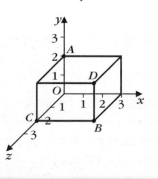

6 Simplify $k + 3k - 2k$

7 Simplify $4 \times a \times b$

8 Simplify $e^2 + e^2 + e^2$

9 (a) Complete the table of values for $y = 3x - 1$.

x	-2	-1	0	1	2	3
y						

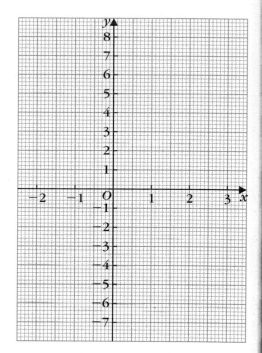

(b) On a copy of the grid, draw the graph of $y = 3x - 1$.
(c) Use your graph to find the value of x when $y = -2.5$

10 Simplify $2p \times 4q$

11 Simplify $g \times g \times g \times g$

12 Expand $3(a - 2b)$

13 Expand and simplify $4(c - 2d) + 2(c + 3d)$

14 Factorise $3w - 12$

15 Factorise $x^2 - 5x$

Check your answers on pages 166–167. For full worked solutions see the CD.

Tick the questions you got right.

Question	1	2ab	2c	3ab	3c	4	5	6	7	8	9	10	11	12	13	14	15
Grade	F	F	D	F	C	C	C	G	F	E	E	D	D	D	C	C	C
Revise this on page	60–61	62–63		60–61		60	63	58	58	59	64–65	58		58	58–59	58–59	58–59	58–59

Mark the grade you are working at on your revision planner on page x.

Go to the pages shown to revise for the ones you got wrong.

Naming and calculating angles

STAGE 1

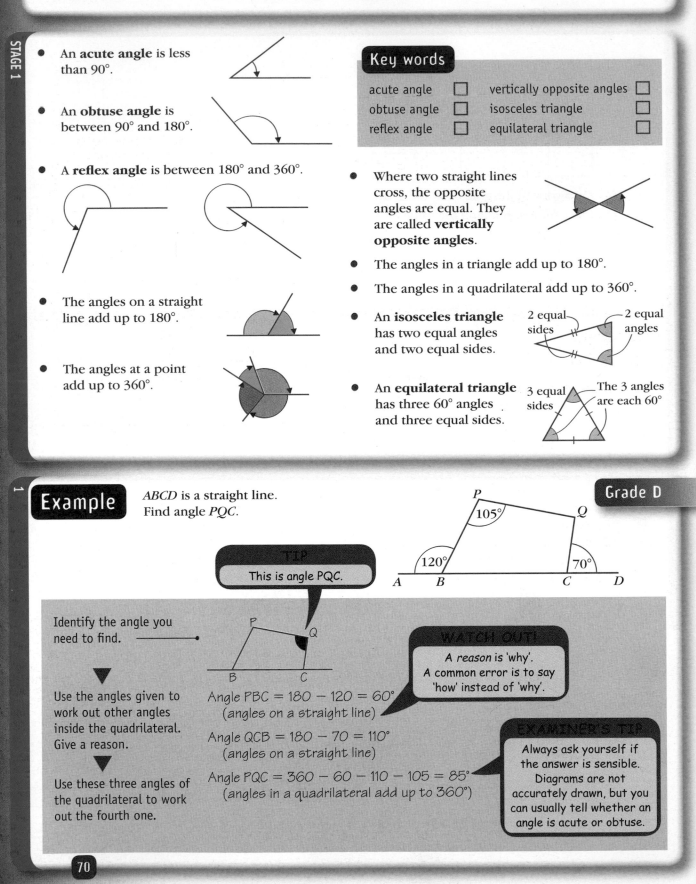

- An **acute angle** is less than 90°.

- An **obtuse angle** is between 90° and 180°.

- A **reflex angle** is between 180° and 360°.

- The angles on a straight line add up to 180°.

- The angles at a point add up to 360°.

Key words

acute angle ☐   vertically opposite angles ☐
obtuse angle ☐   isosceles triangle ☐
reflex angle ☐   equilateral triangle ☐

- Where two straight lines cross, the opposite angles are equal. They are called **vertically opposite angles**.

- The angles in a triangle add up to 180°.

- The angles in a quadrilateral add up to 360°.

- An **isosceles triangle** has two equal angles and two equal sides.

  2 equal sides   2 equal angles

- An **equilateral triangle** has three 60° angles and three equal sides.

  3 equal sides   The 3 angles are each 60°

Example

ABCD is a straight line.
Find angle *PQC*.

Grade D

105°   *P*   *Q*
120°   70°
A   *B*   *C*   *D*

TIP
This is angle PQC.

Identify the angle you need to find.

P   *Q*
B   *C*

WATCH OUT!
A *reason* is 'why'.
A common error is to say 'how' instead of 'why'.

Use the angles given to work out other angles inside the quadrilateral. Give a reason.

Angle PBC = 180 − 120 = 60° (angles on a straight line)

Angle QCB = 180 − 70 = 110° (angles on a straight line)

Use these three angles of the quadrilateral to work out the fourth one.

Angle PQC = 360 − 60 − 110 − 105 = 85° (angles in a quadrilateral add up to 360°)

EXAMINER'S TIP
Always ask yourself if the answer is sensible. Diagrams are not accurately drawn, but you can usually tell whether an angle is acute or obtuse.

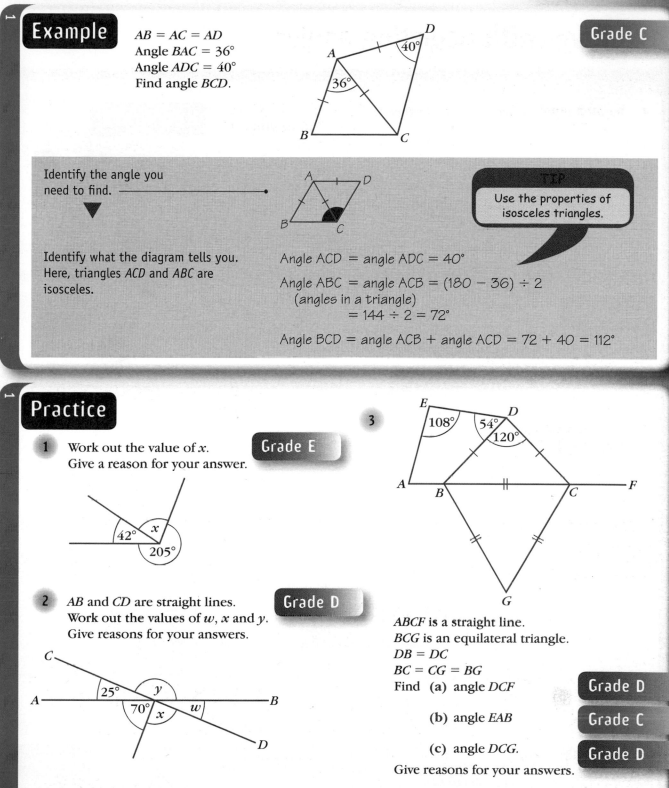

Example

Grade C

$AB = AC = AD$
Angle $BAC = 36°$
Angle $ADC = 40°$
Find angle BCD.

Identify the angle you need to find.

TIP
Use the properties of isosceles triangles.

Identify what the diagram tells you. Here, triangles ACD and ABC are isosceles.

Angle ACD = angle ADC = 40°

Angle ABC = angle ACB = (180 − 36) ÷ 2
(angles in a triangle)
= 144 ÷ 2 = 72°

Angle BCD = angle ACB + angle ACD = 72 + 40 = 112°

Practice

1 Work out the value of x.
Give a reason for your answer.
Grade E

2 AB and CD are straight lines.
Work out the values of w, x and y.
Give reasons for your answers.
Grade D

3

$ABCF$ is a straight line.
BCG is an equilateral triangle.
$DB = DC$
$BC = CG = BG$
Find **(a)** angle DCF Grade D

(b) angle EAB Grade C

(c) angle DCG.   Grade D

Give reasons for your answers.

Check your answers on page 167. For full worked solutions see the CD.
See the Student Book on the CD if you need more help.

Question	1	2	3a	3b	3c
Grade	E	D	D	C	D
Student Book pages	U2 121–122	U2 120–121		U3 178–182	

Working with angles

- **Parallel lines** are shown with arrowheads.

- **Alternate angles** are equal.

- **Corresponding angles** are equal.

Key words

parallel lines ☐
alternate angles ☐
corresponding angles ☐

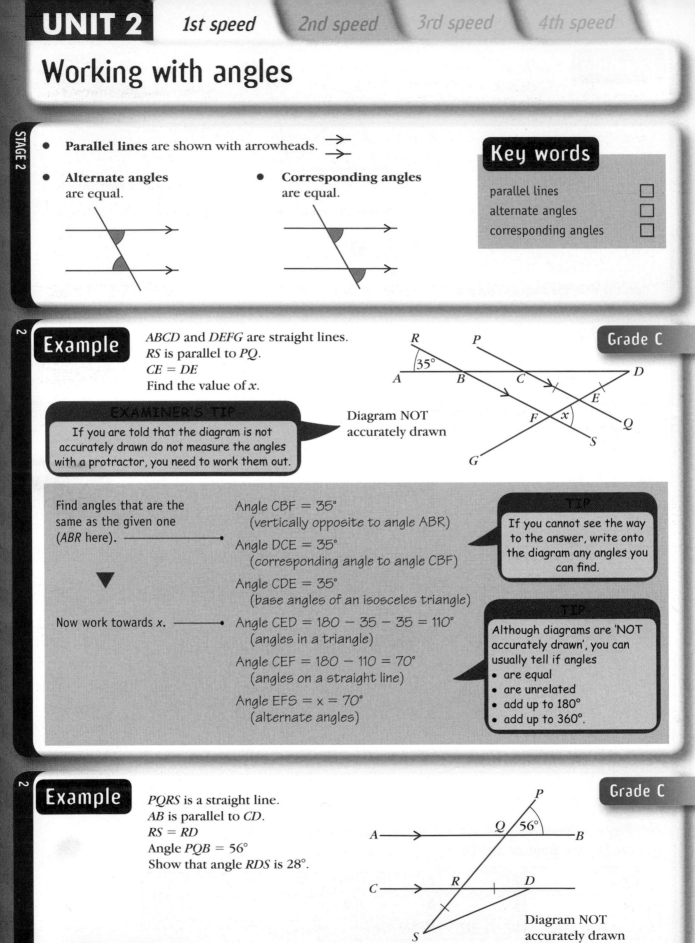

Example

ABCD and *DEFG* are straight lines.
RS is parallel to *PQ*.
CE = *DE*
Find the value of *x*.

Grade C

EXAMINER'S TIP

If you are told that the diagram is not accurately drawn do not measure the angles with a protractor, you need to work them out.

Diagram NOT accurately drawn

Find angles that are the same as the given one (*ABR* here).

▼

Now work towards *x*.

Angle CBF = 35°
(vertically opposite to angle ABR)

Angle DCE = 35°
(corresponding angle to angle CBF)

Angle CDE = 35°
(base angles of an isosceles triangle)

Angle CED = 180 − 35 − 35 = 110°
(angles in a triangle)

Angle CEF = 180 − 110 = 70°
(angles on a straight line)

Angle EFS = x = 70°
(alternate angles)

TIP

If you cannot see the way to the answer, write onto the diagram any angles you can find.

TIP

Although diagrams are 'NOT accurately drawn', you can usually tell if angles
- are equal
- are unrelated
- add up to 180°
- add up to 360°.

Example

PQRS is a straight line.
AB is parallel to *CD*.
RS = *RD*
Angle *PQB* = 56°
Show that angle *RDS* is 28°.

Grade C

Diagram NOT accurately drawn

Find angles that are the same as the given one. ⟶ Angle QRD = 56°
(corresponding angle to angle PQB)

▼

Now work into triangle *SRD*. ⟶ Angle SRD = 180 − 56 = 124°
(angles on a straight line)

▼

The final step. ⟶ Angle RSD = angle RDS = x
(base angles of an isosceles triangle)
x + x + 124 = 180
(angles in a triangle)
So 2x = 56° and x = angle RDS = 28°

> **TIP**
>
> 'Show that' questions are the same as 'Work out' or 'Find' questions except that
> • the answer is given
> • you *must* give all the stages of the working out *and* give reasons.

Practice

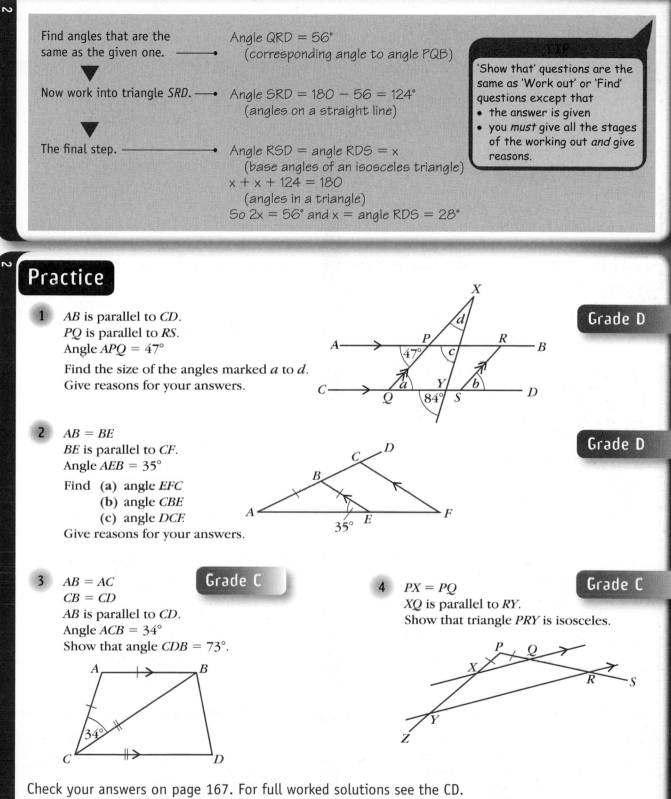

1 *AB* is parallel to *CD*.
PQ is parallel to *RS*.
Angle *APQ* = 47°

Find the size of the angles marked *a* to *d*.
Give reasons for your answers.

Grade D

2 *AB* = *BE*
BE is parallel to *CF*.
Angle *AEB* = 35°

Find **(a)** angle *EFC*
(b) angle *CBE*
(c) angle *DCF*
Give reasons for your answers.

Grade D

3 *AB* = *AC*
CB = *CD*
AB is parallel to *CD*.
Angle *ACB* = 34°
Show that angle *CDB* = 73°.

Grade C

4 *PX* = *PQ*
XQ is parallel to *RY*.
Show that triangle *PRY* is isosceles.

Grade C

Check your answers on page 167. For full worked solutions see the CD.
See the Student Book on the CD if you need more help.

Question	1	2	3	4
Grade	D	D	C	C
Student Book pages	U2 123–124	U2 123–124	U2 124–126	U2 124–126

Angles: topic test

Check how well you know this topic by answering these questions.
First cover the answers on the facing page.

Test questions

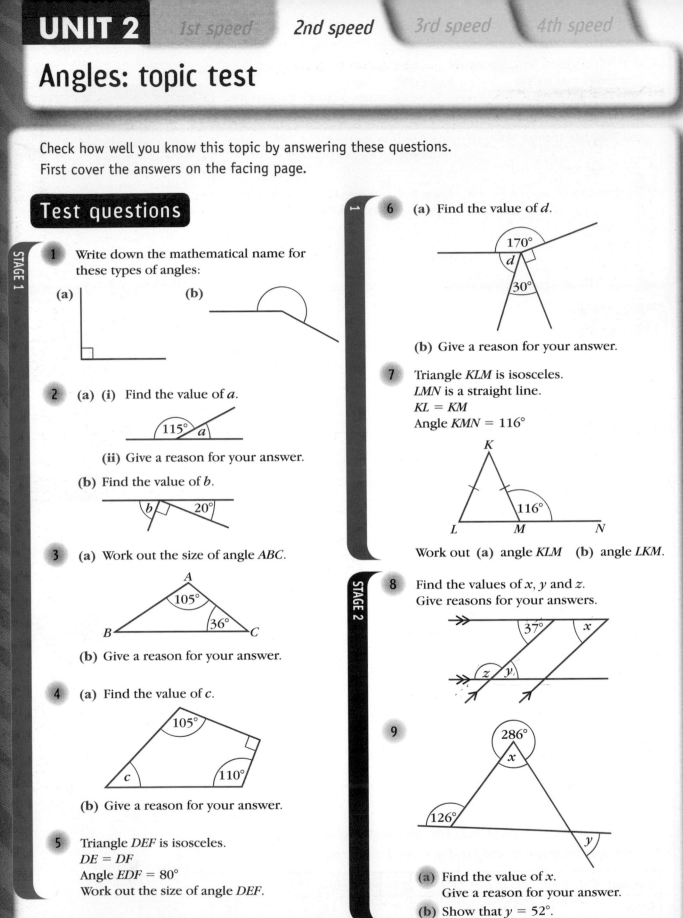

STAGE 1

1   Write down the mathematical name for these types of angles:

(a)   (b)

2   **(a)**   **(i)**   Find the value of a.

     115° a

     (ii)   Give a reason for your answer.

   (b)   Find the value of b.

     b 20°

3   **(a)**   Work out the size of angle ABC.

     A
     105°
     36°
     B C

   (b)   Give a reason for your answer.

4   **(a)**   Find the value of c.

     105°
     c 110°

   (b)   Give a reason for your answer.

5   Triangle DEF is isosceles.
     $DE = DF$
     Angle $EDF = 80°$
     Work out the size of angle DEF.

Now check your answers – see the facing page.

6   **(a)**   Find the value of d.

     170°
     d
     30°

   (b)   Give a reason for your answer.

7   Triangle KLM is isosceles.
     LMN is a straight line.
     $KL = KM$
     Angle $KMN = 116°$

     K
     116°
     L M N

   Work out   **(a)**   angle KLM **(b)**   angle LKM.

STAGE 2

8   Find the values of x, y and z.
     Give reasons for your answers.

     37° x
     z y

9

     286°
     x
     126°
     y

   (a)   Find the value of x.
       Give a reason for your answer.
   (b)   Show that $y = 52°$.

Cover this page while you answer the test questions opposite.

Worked answers

Revise this on...

1 (a) Right angle (b) Reflex angle — page 70 **G**

2 (a) (i) $a = 180 - 115 = 65°$ — page 70 **F**
  (ii) Angles on a straight line add up to 180°.
  (b) $b = 180 - 90 - 20 = 70°$

3 (a) Angle $ABC = 180 - 105 - 36 = 39°$ — page 70 **F**
  (b) Angles in a triangle add up to 180°.

4 (a) $c = 360 - 110 - 105 - 90 = 55°$ — page 70 **E**
  (b) Angles in a quadrilateral add up to 360°.

5 Triangle DEF is isosceles so angle DEF = angle DFE — page 71 **E**
  Angle $DEF = \frac{1}{2}(180 - 80) = 50°$

6 (a) $d = 360 - 30 - 170 - 90 = 70°$ — page 70 **E**
  (b) Angles at a point add up to 360°.

7 (a) Angle $KLM =$ angle $KML = 180 - 116 = 64°$ — page 71 **D**
  (b) Angle $LKM = 180 - 2 \times 64 = 180 - 128 = 52°$

8 $x = 37°$ (corresponding angles) — pages 72–73 **D**
  $y = 37°$ (alternate angles)
  $z = 180 - 37 = 143°$ (angles on a straight line)

9 (a) $x = 360 - 286 = 74°$ (angles at a point) — pages 72–73 **E**

  (b) Angle alongside $126° = 180 - 126 = 54°$ (angles on a straight line) **C**
  Third angle in the triangle $= 180 - 74 - 54 = 52°$
  $y = 52°$ (vertically opposite angles)

Tick the questions you got right.

Question	1	2	3	4	5	6	7	8	9a	9b
Grade	G	F	F	E	E	E	D	D	E	C

Mark the grade you are working at on your revision planner on page x.

Units of measurement

STAGE 2

2

- Approximate **conversions** between units:

Metric	Imperial
8 km	5 miles
1 kg	2.2 pounds
25 g	1 ounce
1 l	$1\frac{3}{4}$ pints
4.5 l	1 gallon
1 m	39 inches
30 cm	1 foot
2.5 cm	1 inch

- **Speed** $= \dfrac{\text{distance}}{\text{time}}$

- **Average speed** $= \dfrac{\text{total distance}}{\text{total time}}$

- Units of speed are miles per hour (mph), kilometres per hour (km/h) and metres per second (m/s).

- If you make a measurement correct to a given unit the true value lies in a range that extends half a unit below and half a unit above the measurement.

Key words

- metric ☐
- imperial ☐
- conversion ☐
- speed ☐
- average speed ☐
- timetable ☐

Example   Change 50 litres into gallons.   Grade E

Use the unitary method. ———→   4.5 l is 1 gallon

1 l is $\dfrac{1}{4.5}$ gallons

50 l is $50 \times \dfrac{1}{4.5} = 11.1$ gallons.

TIP
Gallons are larger than litres so there are less of them. So you *divide* by 4.5 to turn litres into gallons.

Example   Grade F

Here is part of a railway **timetable**.

(a) How long does it take the 08 27 train from Lincoln to get to Sheffield?

(b) How long does the last train take to travel from Retford to Worksop?

(c) Gillian lives in Retford. She has to be in Sheffield by quarter to eleven. What is the time of the latest train she can catch?

	First train			*Last train*
Lincoln	07 04	08 27	09 27	17 16
Retford	07 40	09 03	10 03	17 52
Worksop	07 54	09 15	10 15	18 04
Shireoaks	07 57	09 18	—	—
Sheffield	08 26	09 47	10 46	18 26

STAGE 1 AND 2

Find the column for the 08 27 train. Go down to the Sheffield row.

(a) The 08 27 gets to Sheffield at 09 47
09 47 − 08 27 = 1 hour 20 minutes

WATCH OUT!
A common error is to use a calculator in its decimal mode.

Use the times in the last column.

(b) The last train leaves Retford at 17 52 and arrives in Worksop at 18 04. This is 8 minutes before 18 00 and 4 minutes after. It takes 12 minutes.

TIP
It is best to work 'in pieces' here and add them up.

(c) Quarter to eleven is 10 45 so the third train is too late. She must be at the station before 09 03 to catch the second train.

Example

Bianca cycles 3600 metres in 12 minutes.
Work out her average speed in kilometres per hour (km/h).

Grade C

Change to the units you want.

$12 \text{ minutes} = \frac{12}{60} = \frac{1}{5} \text{ hour}$

$3600 \text{ metres} = 3.6 \text{ kilometres}$

TIP

The speed is in km/h, so work in hours. (If the speed is in m/s, work in seconds.)

WATCH OUT!

12 minutes is not 0.12 hour.

$\text{Speed} = \frac{\text{distance}}{\text{time}}$

$= \frac{3.6 \text{ km}}{\frac{1}{5} \text{ hour}} = 3.6 \times 5 = 18 \text{ km/h}$

Practice

STAGE 1 AND 2

1 Use the railway timetable in the example on page 76 to answer these questions.

(a) At what time does the second train get to Worksop?

Grade G

(b) How long does the last train take to get from Retford to Sheffield?

Grade G

(c) Ravi takes 15 minutes to walk from home to the Lincoln station.
He allows 5 minutes to buy a ticket and get on to the platform.
What is the latest time he can leave home to catch the first train?

Grade F

2 Write down the readings on these scales.

(a)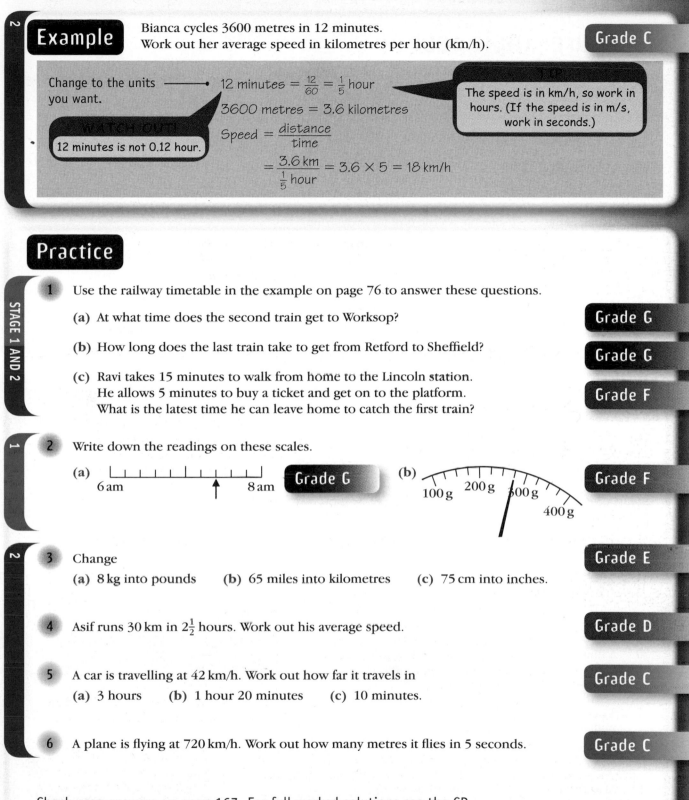

6 am   8 am

Grade G

(b)

100 g   200 g   300 g

400 g

Grade F

3 Change

Grade E

(a) 8 kg into pounds   (b) 65 miles into kilometres   (c) 75 cm into inches.

4 Asif runs 30 km in $2\frac{1}{2}$ hours. Work out his average speed.

Grade D

5 A car is travelling at 42 km/h. Work out how far it travels in

Grade C

(a) 3 hours   (b) 1 hour 20 minutes   (c) 10 minutes.

6 A plane is flying at 720 km/h. Work out how many metres it flies in 5 seconds.

Grade C

Check your answers on page 167. For full worked solutions see the CD.
See the Student Book on the CD if you need more help.

Question	1ab	1c	2a	2b	3	4	5	6
Grade	G	F	G	F	E	D	C	C
Student Book pages	U2 134–135		U2 136–137		U2 141–142	U2 144–145	U2 144–145	U2 146–147

Measure: topic test

Check how well you know this topic by answering these questions.
First cover the answers on the facing page.

Test questions

STAGE 2

1 Which is the most sensible metric unit for measuring

     (a) the length of a pencil   **(b)** the weight of an exercise book

     (c) the capacity of a cup?

2 Write

     (a) 5.6 kilograms in grams   **(b)** 2360 centimetres in metres   **(c)** 750 m*l* in litres.

3 Change these to 24-hour clock times.

     (a) 3:25 am   **(b)** 10:50 am   **(c)** 2:35 pm   **(d)** quarter to eleven in the evening

4 Change these to 12-hour clock times (am or pm).

     (a) 07:45   **(b)** 10:30   **(c)** 17:25   **(d)** 22:40   **(e)** 12:05

STAGE 1

5 For each scale write down what one small division is worth and write down the measurement shown by the arrow.

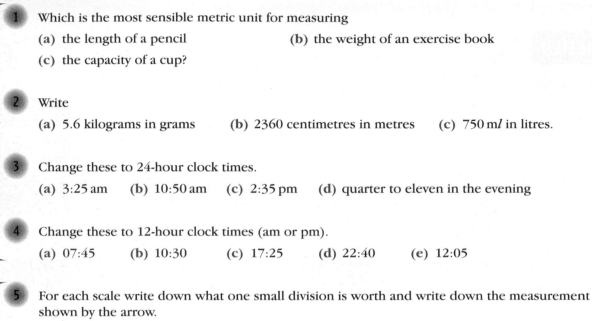

2

6 A film starts at 18:40 and finishes at 21:15.
Work out the running time in hours and minutes.

7 14 pounds = 1 stone
William weighs 8 stone 3 pounds.
Work out William's weight in kilograms.

8 Alan runs 20 km in 1 hour 40 minutes.
Work out his average speed in km/h.

9 Pia drives the 95 kilometres from London to Brighton at an average speed of 42 km/h.
How long does she take?

Now check your answers – see the facing page.

Cover this page while you answer the test questions opposite.

Worked answers

Revise this on...

1 (a) centimetres   (b) grams   (c) millilitres   **G**

2 (a) 5600 g   (b) 23.6 m   (c) 0.75 *l*   **G**

3 (a) 03:25   (b) 10:50   (c) 14:35   (d) 22:45   **G**

4 (a) 7:45 am   (b) 10:30 am   (c) 5:25 pm   **G**
   (d) 10:40 pm   (e) 12:05 pm

5 (a) 0.1, 1.3   (b) 25 g, 175 g   page 77   **G**

   (c) 6 minutes, 10:42   (d) 20, 360   **F**

6 From 18 40 to 19 00 is 20 minutes   page 76   **E**
   From 19 00 to 21 00 is 2 hours
   From 21 00 to 21 15 is 15 minutes
   Total running time = 2 hours 35 minutes

7 8 stones = 8 × 14 = 112 pounds   page 76   **E**
   William weighs 112 + 3 = 115 pounds = 115 ÷ 2.2 kg = 52.27 kg

8 Average speed = $20 \div 1$ hour 40 minutes = $20 \div 1\frac{2}{3} = 20 \div \frac{5}{3}$ page 77   **D**
   $\qquad\qquad\qquad\qquad\qquad\qquad\qquad = 20 \times \frac{3}{5}$
   $\qquad\qquad\qquad\qquad\qquad\qquad\qquad = 12$ km/h

9 Time = 95 ÷ 42 = 2.2619 hours   page 77   **D**
   0.2619 hours = 0.2619 × 60 = 16 minutes
   Jane takes 2 hours 16 minutes.

Tick the questions you got right.

Question	1	2	3	4	5ab	5cd	6	7	8	9	10
Grade	G	G	G	G	G	F	E	E	D	D	C

Mark the grade you are working at on your revision planner on page x.

Perimeter and area

STAGE 1

- The **perimeter** of a 2-D shape is the distance around the edge of the shape.

- The **area** of a 2-D shape is a measure of the amount of space it covers. Typical units of area are mm², cm², m² and km².

Key words

perimeter	☐	triangle	☐
area	☐	parallelogram	☐
rectangle	☐	trapezium	☐

- Area of a **rectangle**
  = length × width
  = $l \times w$

- Area of a **parallelogram**
  = base × vertical height = $b \times h$

- Area of a **triangle**
  = $\frac{1}{2}$ base × height
  = $\frac{1}{2} \times b \times h$

- Area of a **trapezium**
  = $\frac{1}{2}$ × sum of parallel sides × distance between parallels
  = $\frac{1}{2}(a + b) \times h$

1

Example
Find the area of this shape.

TIP

Grade D

Decide how to split up the diagram. Here, it is best to complete the rectangle and then subtract the area of the triangle.

Area of rectangle = 12 × 6 = 72 cm²
Base of triangle = 12 − 4 − 2 = 6 cm
Area of triangle = $\frac{1}{2}$ × 6 × 3 = 9 cm²
Area of shape = 72 − 9 = 63 cm²

1

Example
Work out the area of this shape.

Grade D

Use the formula $\frac{1}{2}(a + b) \times h$

Area of trapezium = $\frac{1}{2}(4 + 7) \times 6 = \frac{1}{2} \times 11 \times 6 = \frac{1}{2} \times 66 = 33$ cm²

STAGE 2

- The **surface area** is the total area of all the faces of a solid shape.
- A **prism** is a shape which has a uniform cross-section.

Key words

surface area ☐   prism ☐

Example

Find the total surface area of this prism.

TIP

Imagine 'unwrapping' the four rectangles to make one big rectangle. Its length is the same as the perimeter of the end.

WATCH OUT!

Don't forget the base.

The prism has six faces: two ends and the four rectangles wrapping round these ends.

Area of one end $= \frac{1}{2}(6 + 9) \times 4 = 30\,cm^2$
Perimeter of the trapezium $= 4 + 6 + 5 + 9 = 24\,cm$
Total area of surrounding rectangles $= 24 \times 8 = 192\,cm^2$
Total surface area $= 30 + 30 + 192 = 252\,cm^2$

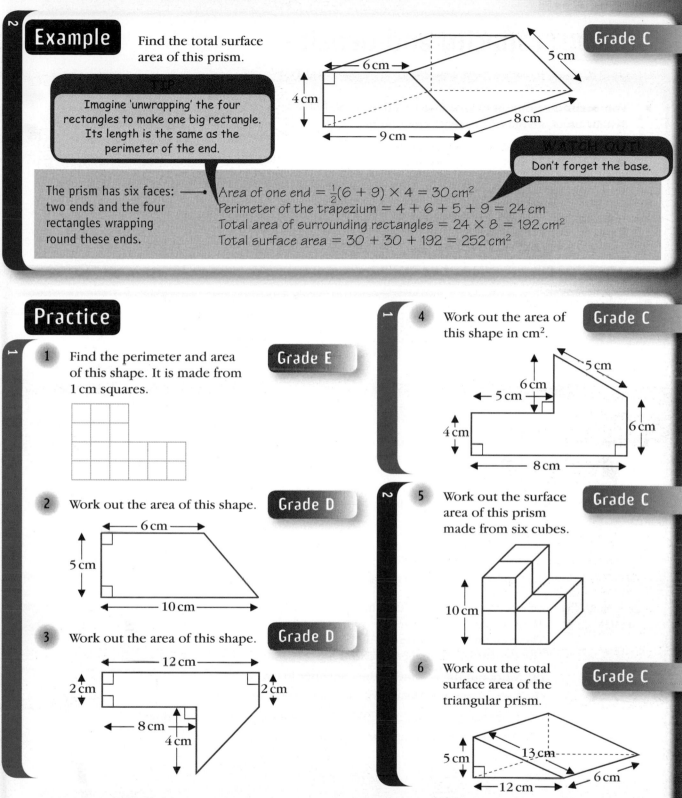

Practice

1 Find the perimeter and area of this shape. It is made from 1 cm squares.

2 Work out the area of this shape.

6 cm
5 cm
10 cm

3 Work out the area of this shape.

12 cm
2 cm   2 cm
8 cm
4 cm

4 Work out the area of this shape in cm².

5 cm
6 cm
5 cm
6 cm
4 cm
8 cm

5 Work out the surface area of this prism made from six cubes.

10 cm

6 Work out the total surface area of the triangular prism.

5 cm   13 cm
12 cm   6 cm

Check your answers on page 167. For full worked solutions see the CD.
See the Student Book on the CD if you need more help.

Question	1	2	3	4	5	6
Grade	E	D	D	C	C	C
Student Book pages	U2 155–157	U2 159–161	U2 159–161	U2 159–161	U2 161–162	U2 157–162

Volume, capacity and density

STAGE 2

- **Volume** is the amount of space occupied by a solid object. Typical units of volume are mm^3, cm^3 and m^3.

- Volume of a **cuboid**
  = length × width × height = $l \times w \times h$

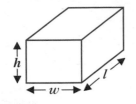

Key words

volume	☐
cuboid	☐
prism	☐
cylinder	☐
capacity	☐
density	☐

- Volume of a **prism**
  = area of cross-section × length

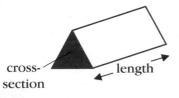

- Volume of a **cylinder**
  = area of cross-section × length = $\pi r^2 h$

> **TIP**
> For areas of circles see page 142.

- **Capacity** is the amount a container can hold. Typical units of capacity are millilitres (ml), centilitres (cl) and litres (l).

- In terms of space required, $1\,ml = 1\,cm^3$

- **Density** $= \dfrac{\text{mass}}{\text{volume}}$
  Density is measured in g/cm^3 or kg/m^3.

2

Example

An open container is in the shape of a cuboid measuring 30 cm by 20 cm by 20 cm.
The container is completely full of orange juice.
The density of the orange juice is $1.1\,g/cm^3$.

(a) Work out the volume of the container in cm^3.

(b) Write down the capacity of the container in litres.

(c) How many cups holding 300 ml can be filled from the container?

(d) Work out the mass of orange juice in each cup.

Grade C

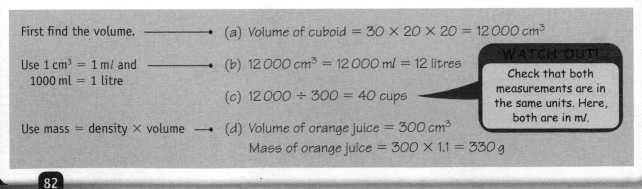

First find the volume. ──────• (a) Volume of cuboid = 30 × 20 × 20 = 12 000 cm³

Use 1 cm³ = 1 ml and ──────• (b) 12 000 cm³ = 12 000 ml = 12 litres
1000 ml = 1 litre

(c) 12 000 ÷ 300 = 40 cups

> **WATCH OUT!**
> Check that both measurements are in the same units. Here, both are in ml.

Use mass = density × volume ──• (d) Volume of orange juice = 300 cm³
Mass of orange juice = 300 × 1.1 = 330 g

Practice

1 Work out the capacity of a large box in the shape of a cuboid. **Grade E**

The inside of the box measures 100 cm by 60 cm by 45 cm.

Give your answer in litres.

2 Work out the volume of this cuboid. **Grade E**

3 cm
20 cm
8 cm

3 Work out the volume of this shelf in cm³. **Grade D**

8 cm
4 mm
0.6 m

4 The volume of this cuboid is 420 cm³. **Grade D**
Work out its height.

10 cm
6 cm

5 (a) Work out the volume of this prism. **Grade D**

4 cm
4 cm
8 cm
6 cm

(b) The prism is made of glass with density 2.6 g/cm³. **Grade C**
Work out the weight of the prism.

6 An ingot of gold has a mass of 4 kg. **Grade C**
The volume of the ingot is 210 cm³.
Calculate the density of the gold

(a) in g/cm³

(b) in kg/m³

7 Find the volume of a cylinder with **Grade C**

(a) length 6 cm, radius 10 cm

(b) diameter 14 cm, height 20 cm.

Check your answers on page 167. For full worked solutions see the CD.
See the Student Book on the CD if you need more help.

Question	1	2	3	4	5a	5b	6	7
Grade	E	E	D	D	D	C	C	C
Student Book pages	U2 163–165	U2 163–165	U2 163–165	U2 163–165	U2 165–167	U2 148–149	U2 148–149	U2 165–167

Perimeter, area and volume: topic test

Check how well you know this topic by answering these questions.
First cover the answers on the facing page.

Test questions

STAGE 1

1   Work out the perimeter and area of this shape.

2   The area of this rectangle is 30 cm².
Its length is 10 cm.
Work out its height.

30 cm²

← 10 cm →

3   Work out the area of this triangle.

8 cm   17 cm

← 15 cm →

STAGE 2

4   Packets measuring 5 cm by 3 cm by 10 cm
are packed into a box which measures
50 cm by 60 cm by 30 cm.
Work out how many packets will exactly
fill the box.

5   Work out the volume of

(a) a box which measures 15 cm by 10 cm
by 8 cm

(b) a plank of wood which measures 10 cm
by 25 mm by 2 m.

6   The cross-section of a plank of wood is 8 cm
by 40 mm. Its volume is 5760 cm³.
Work out the length of the plank in metres.

7   Work out the area of this shape.

3 cm

10 cm

6 cm

← 6 cm →

8   The face *ABCD* of this prism is a trapezium.
AB = 14 cm
AC = 12 cm
CD = 23 cm
Angle *ACD* = 90°
The length is 40 cm.

(a) Work out the volume of the prism.

(b) Work out the total surface area.

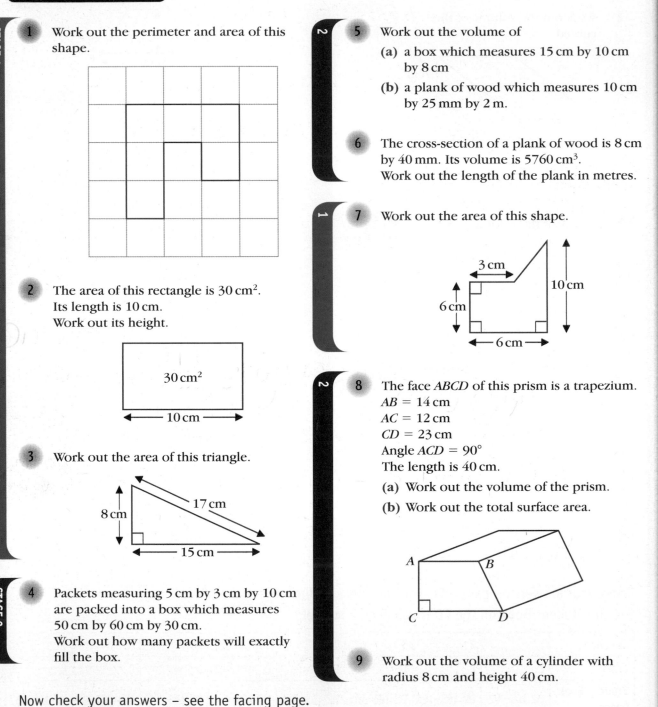

9   Work out the volume of a cylinder with
radius 8 cm and height 40 cm.

Now check your answers – see the facing page.

Cover this page while you answer the test questions opposite.

Worked answers

Revise this on...

1   Perimeter = 3 + 2 + 1 + 1 + 1 + 2 + 1 + 3 = 14 cm
    Area = number of squares = 6 cm^2 page 80   G

2   Area = height × 10 = 30 cm^2 page 80   F
    Height = 3 cm

3   Area = $\frac{1}{2}$ base × height = $\frac{1}{2}$ × 15 × 8 = 60 cm^2 page 80   E

4   The 50 cm side will take a row of ten 5 cm sides. page 82   E
    The 60 cm side will take a layer of 20 rows using the 3 cm side.
    The 30 cm side (height) will take 3 layers of 10 cm
    Total number of packets = 10 × 20 × 3 = 600

5   (a) Volume = 15 × 10 × 8 = 1200 cm^3 page 82   D

    (b) Change lengths to centimetres first: 25 mm = 2.5 cm and 2 m = 200 cm
         Volume = 10 × 2.5 × 200 = 5000 cm^3

6   40 mm = 4 cm. Cross-section = 4 × 8 = 32 cm^2 page 82   D
    Length = 5760 ÷ 32 = 180 cm = 1.8 metres

7   Height of triangle = 10 − 6 = 4 cm page 80   C
    Base of triangle = 6 − 3 = 3 cm
    Area = 6 × 6 + $\frac{1}{2}$ × 3 × 4 = 36 + 6 = 42 cm^2

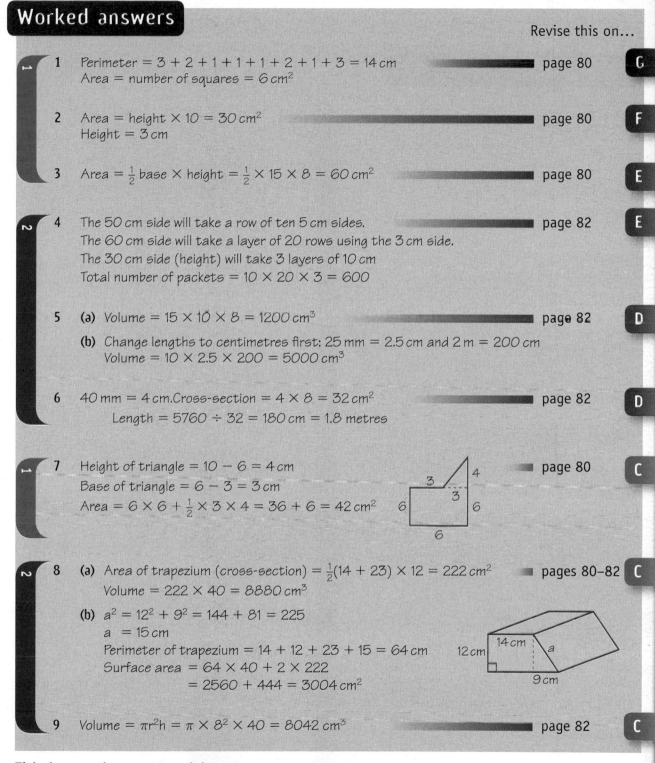

8   (a) Area of trapezium (cross-section) = $\frac{1}{2}$(14 + 23) × 12 = 222 cm^2 pages 80–82   C
         Volume = 222 × 40 = 8880 cm^3

    (b) a^2 = 12^2 + 9^2 = 144 + 81 = 225
         a = 15 cm
         Perimeter of trapezium = 14 + 12 + 23 + 15 = 64 cm
         Surface area = 64 × 40 + 2 × 222
                      = 2560 + 444 = 3004 cm^2

9   Volume = $\pi r^2 h$ = π × 8^2 × 40 = 8042 cm^3 page 82   C

Tick the questions you got right.

Question	1	2	3	4	5	6	7	8	9
Grade	G	F	E	E	D	D	C	C	C

Mark the grade you are working at on your revision planner on page x.

Shape, space and measure: subject test

Exam practice questions

1 Work out the perimeter and area of these shapes.

(a)

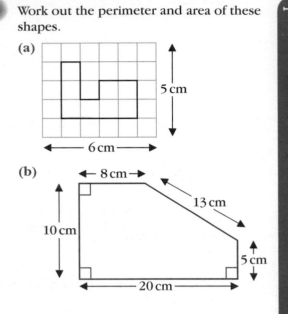

5 cm

6 cm

(b)

8 cm

13 cm

10 cm

5 cm

20 cm

2 In the diagram, *ABC* is an equilateral triangle. *BDC* is an isosceles triangle with *DB* = *BC*. Angle *BDC* = 110°.

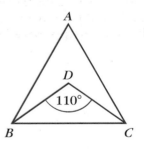

A

D

110°

B   C

Find angle *ABD*.

3 Work out the area of the shapes shown.

(a)

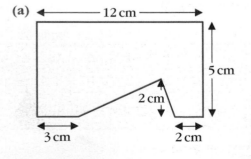

12 cm

5 cm

2 cm

3 cm   2 cm

(b)

4 cm

3 cm

5 cm

4 cm

4 cm

(c)

8 cm

2 cm

6 cm

4 *ABDE* is a square. *BCD* is an equilateral triangle.

Work out

(a) angle *ABC*

(b) angle *ACD*

Give reasons for your answers.

A   B

C

E   D

5 Work out the size of the angles marked with letters.

Give reasons for your answers.

(a)

100°

a   34°

(b)

30°   c

(c)

43° 110°

x   y

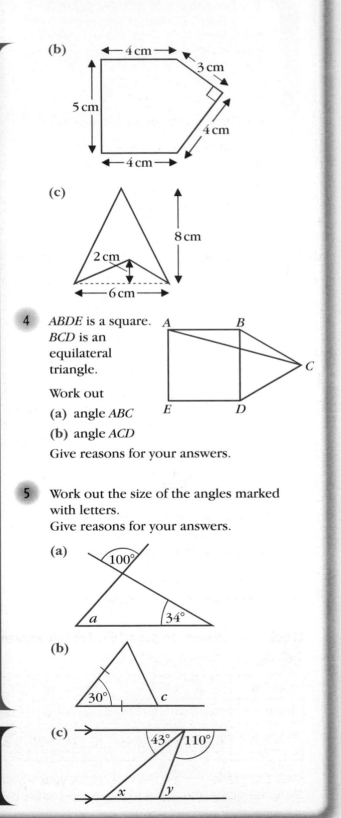

STAGE 2

6 Work out the volume of ^ this prism by counting cubes.

7 Peter drives 225 kilometres in 4 hours 30 minutes. Work out his average speed for this journey.

8 How long does it take to run

(a) 400 metres at an average speed of 8 m/s

(b) 1500 metres at 15 km/h?

9 The radius of a cylinder is 5 cm and the height is 15 cm.

Give units with your answer°.

10

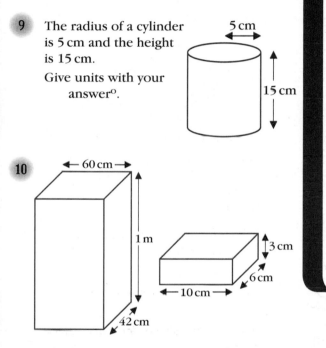

The larger carton measures
1 metre × 60 cm × 42 cm
The smaller box measures
3 cm × 10 cm × 6 cm
Work out how many of the smaller boxes are needed to totally fill the carton.

11 Find **(a)** angle *BRY*, **(b)** angle *ABZ*, **(c)** angle *WQP*, **(d)** angle *RBC*
Give reasons for your answers.

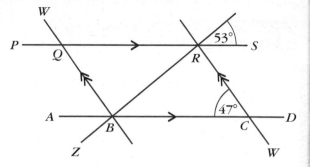

12 Work out the surface area of this shape.

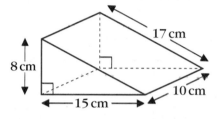

13 Sheila cycles at an average speed of 20 km/h for 2 hours 15 minutes.
How far does she cycle?

14 A block of wood is in the shape of a cuboid.
It measures 20 cm by 8 cm by 6 cm.
The density of the wood is 0.8 g/cm³.
Work out the weight of the block.

Check your answers on page 167. For full worked solutions see the CD.
Tick the questions you got right.

Question	1	2	3	4	5ab	5c	6	7	8a	8b	9	10	11	12	13	14
Grade	G	D	D	C	D	D	G	E	E	D	D	D	D	C	C	C
Revise this on page	80	70	80	70	70–71	72–73	82	76–77	76–77		82	82–83	72–73	82	76–77	82

Mark the grade you are working at on your revision planner on page x.
Go to the pages shown to revise for the ones you got wrong.

Unit 2 Key points

Number

STAGE 1

Integers

- To write a number to the **nearest 10,** look at the **units digit**.

- To write a number to the **nearest 100**, look at the **tens digit**.

  (If it is 5 or more, round up. If it is less than 5, round down.)

- To write a number to the **nearest 1000**, look at the **hundreds digit**.

- Adding a negative number has the same effect as subtracting the positive number:

  $4 + -1 = 3$

- Subtracting a negative number has the same effect as adding the positive number:

  $2 - -3 = 5$

- When multiplying or dividing two **like signs** give a $+$, two **unlike signs** give a $-$

STAGE 2

- The **power** is how many times a number is multiplied by itself: $2 \times 2 \times 2 \times 2 = 2^4$. A power is also called an **index** (plural **indices**).

- To **multiply** powers of the same number, add the indices:

  $2^3 \times 2^4 = 2^{3+4} = 2^7$

- To **divide** powers of the same number, subtract the indices:

  $5^6 \div 5^4 = 5^{6-4} = 5^2$

- **BIDMAS** is a made-up word to help you remember the order of operations:

B I D M A S

Brackets   Indices   Divide   Multiply   Add   Subtract

1

- The **factors** of a number are whole numbers that divide exactly into the number.
  1, 2, 3, 4, 6 and 12 are factors of 12

- **Multiples** of a number are the results of multiplying the number by a positive whole number.
  3, 6 and 9 are multiples of 3

- A **prime number** is a number with only 2 factors, 1 and itself.
  1 is not a prime number as it can only be divided by one number (itself).

- To **estimate** the approximate answer to a calculation, round each number to
  1 significant figure (1 s.f.).

Decimals, fractions and percentages

- When you multiply decimals, work out the multiplication without the decimal points and put in the decimal point at the end.

- When dividing by decimals make sure you always divide by a whole number by multiplying both numbers by 10, 100 or 1000 etc. Make sure the decimal point in the answer lines up with the one in the question.

- To round to a given number of **decimal places (d.p.)**, count the number of decimal places and look at the next digit. If it is 5 or more, round up; if it is less than 5, round down.

- To round to a given number of **significant figures** (s.f.), count the number of digits from the first non-zero digit, starting from the *left*. Look at the next digit to decide whether to round up or down.

- Top heavy fractions are called **improper fractions**.

- An improper fraction can be written as a **mixed number**, $\frac{12}{5} = 2\frac{2}{5}$.

- **Equivalent fractions** have the same value and can be cancelled down into their simplest terms:
  $$\frac{8}{12} = \frac{4}{6} = \frac{2}{3}.$$

Algebra

Algebra, sequences and line graphs

- An **algebraic expression** is a collection of letters, symbols and numbers:
  $$a + 3b - 2c$$

- You can combine **like terms** by adding and subtracting them:
  $$2a + 3a = 5a \qquad \text{and} \qquad 3b + 4b - b = 6b$$

- You can simplify algebraic expressions by collecting like terms together:
  $$2a - 4b + 3a + 5b \qquad \text{simplifies to} \qquad 5a + b$$

- **Expanding** the brackets means multiplying to remove the brackets:
  $$4(3a + b) = 12a + 4b$$

- **Factorising** means splitting up an expression using brackets:
  $$12a + 4b = 4(3a + b)$$

- In a number pattern or **sequence** there is always a **rule** to get from one number to the next.

- To find the rule for the nth term of a number pattern, use a table of values.

Term number	1	2	3	4
Term	5	8	11	14
Difference		+3	+3	+3

The rule is $3n + 2$.

- The number line is **1-dimensional** or **1-D**. You can describe positions on the number line using one number or coordinate, for example (2).

- Flat shapes are **2-dimensional** or **2-D**. You can describe positions on a flat shape using two numbers or coordinates, for example (2, 1).

- Solid shapes are **3-dimensional** or **3-D**. You can describe positions in a solid shape using three numbers or coordinates, for example (4, 1, 2).

- The **equation of a line** uses algebra to show a relationship between the x- and y-coordinates of points on the line.

- An equation containing an x-term and no higher powers of x (such as x^2) is a **linear equation**.

- The graph of a linear equation is a straight line.

Shape, space and measure

Angles

- The angles on a straight line add up to 180°.

- The angles at a point add up to 360°.

- The angles in a triangle add up to 180°.

- The angles in a quadrilateral add up to 360°.

- **Alternate angles** are equal.

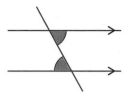

- **Corresponding angles** are equal.

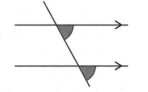

Measure

- **Average speed** $= \dfrac{\text{total distance}}{\text{total time}}$ (take care with units)

  8 kilometres = 5 miles
  1 kilogram = 2.2 pounds

Perimeter, area and volume

- Area of a triangle $= \frac{1}{2}$ base $\times$ height $= \frac{1}{2} \times b \times h$

- Area of a parallelogram $=$ base $\times$ vertical height $= b \times h$

- Area of a trapezium $= \frac{1}{2} \times$ sum of parallel sides $\times$ distance between parallels $= \frac{1}{2}(a + b) \times h$

- The **surface area** is the total area of all the faces of a solid shape.

- **Volume** is the amount of space occupied by a solid object. Typical units of volume are mm^3, cm^3 and m^3.

- Volume of a cuboid $=$ length $\times$ width $\times$ height $= l \times w \times h$

- **Capacity** is the amount a container can hold. Typical units of capacity are millilitres (ml), centilitres (cl) and litres (l).

- In terms of space required, $1\,ml = 1\,\text{cm}^3$

Unit 2 Examination practice paper

A formula sheet can be found on page 161.

Stage 1 (multiple choice)

Non-calculator

1 A golf tournament had 5678 spectators.

Which of the numbers below shows the number 5678 to the nearest ten?

A 5700   **B** 5678   **C** 5670

D 5800   **E** 5680

2 Here is part of a train timetable.

Manchester	08 30	09 00	09 40
Levenshulme	08 38	09 08	09 48
Heaton Chapel	08 42	09 12	09 52
Stockport	08 47	09 17	09 57

When will the 09 40 train from Manchester arrive at Heaton Chapel?

A 09 48   **B** 09 12   **C** 09 17

D 09 52   **E** 09 57

3   61   47   72   53   32

If these numbers are put in order from lowest to highest number, which would be the middle number?

A 61   **B** 47   **C** 72   **D** 53   **E** 32

4 What type of angle is this?

A Acute

B Obtuse

C Reflex

D Right-angled   **E** Opposite

5 What is the value of the 4 in the number 24 752?

A 4   **B** 40   **C** 400

D 4000   **E** 40 000

6 What is the reading on the number line?

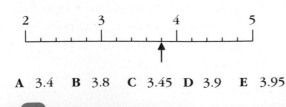

A 3.4   **B** 3.8   **C** 3.45   **D** 3.9   **E** 3.95

7 This is a series of patterns made from dots.

Pattern 1   Pattern 2   Pattern 3

Pattern number	1	2	3
Number of dots	6	9	12

How many dots are there in Pattern 4?

A 14   **B** 15   **C** 16   **D** 17   **E** 18

8 What is the size of the angle marked x.

Diagram **NOT** accurately drawn

70° x 60°   30°

A 10°   **B** 20°   **C** 30°   **D** 40°   **E** 50°

9   2°C   −5°C   3°C   −8°C   6°C

If these temperatures are written in order with the lowest temperature first, which temperature will be 4th in order?

A 2°C   **B** −5°C   **C** 3°C   **D** −8°C   **E** 6°C

10 Here are the first five terms in a sequence of numbers.

7   11   15   19   23

Which is the 9th term in this sequence?

A 27   **B** 35   **C** 38   **D** 39   **E** 45

11

What are the coordinates of the point P?

A $(4, -2)$ **B** $(2, 4)$ **C** $(-2, 4)$

D $(4, 2)$ **E** $(-2, -4)$

12 Work out $28 - 4 \times 5$

    A 8   **B** 12   **C** 100   **D** 120   **E** 125

13 What is 0.634 written as a fraction?

    A $6\frac{34}{100}$ **B** $\frac{634}{10}$ **C** $\frac{634}{100}$

    D $\frac{634}{1000}$ **E** $\frac{634}{10000}$

14

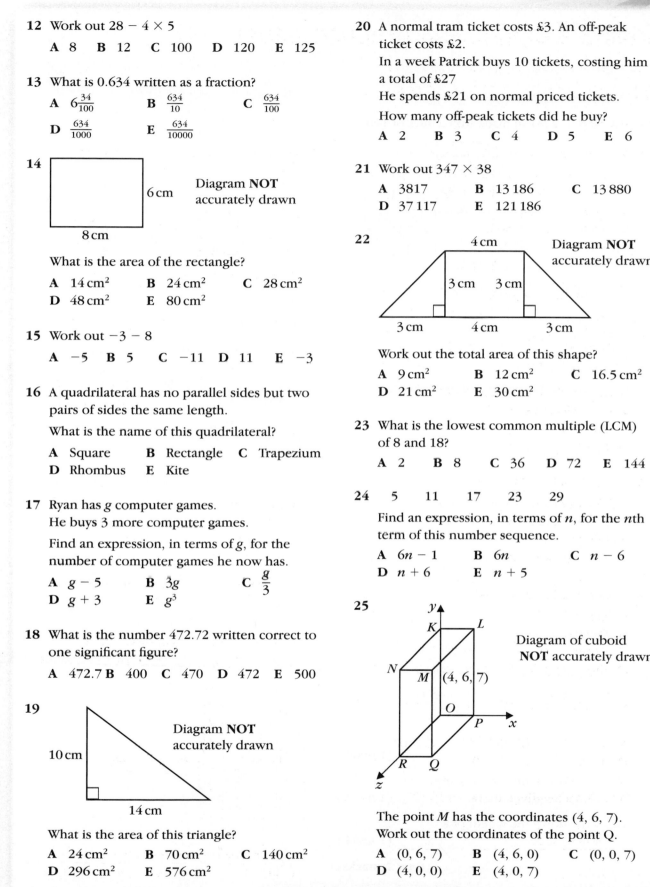

6 cm

8 cm

Diagram **NOT** accurately drawn

What is the area of the rectangle?

    A 14 cm²   **B** 24 cm²   **C** 28 cm²

    D 48 cm²   **E** 80 cm²

15 Work out $-3 - 8$

    A -5 **B** 5   **C** -11 **D** 11   **E** -3

16 A quadrilateral has no parallel sides but two pairs of sides the same length.

What is the name of this quadrilateral?

    A Square   **B** Rectangle   **C** Trapezium

    D Rhombus   **E** Kite

17 Ryan has g computer games.

He buys 3 more computer games.

Find an expression, in terms of g, for the number of computer games he now has.

    A $g - 5$ **B** $3g$ **C** $\frac{g}{3}$

    D $g + 3$ **E** g^3

18 What is the number 472.72 written correct to one significant figure?

    A 472.7   **B** 400   **C** 470   **D** 472   **E** 500

19

Diagram **NOT** accurately drawn

10 cm

14 cm

What is the area of this triangle?

    A 24 cm²   **B** 70 cm²   **C** 140 cm²

    D 296 cm²   **E** 576 cm²

20 A normal tram ticket costs £3. An off-peak ticket costs £2.

In a week Patrick buys 10 tickets, costing him a total of £27

He spends £21 on normal priced tickets.

How many off-peak tickets did he buy?

    A 2   **B** 3   **C** 4   **D** 5   **E** 6

21 Work out 347×38

    A 3817   **B** 13 186   **C** 13 880

    D 37 117   **E** 121 186

22

4 cm

3 cm   3 cm

3 cm   4 cm   3 cm

Diagram **NOT** accurately drawn

Work out the total area of this shape?

    A 9 cm²   **B** 12 cm²   **C** 16.5 cm²

    D 21 cm²   **E** 30 cm²

23 What is the lowest common multiple (LCM) of 8 and 18?

    A 2   **B** 8   **C** 36   **D** 72   **E** 144

24   5   11   17   23   29

Find an expression, in terms of n, for the nth term of this number sequence.

    A $6n - 1$ **B** $6n$ **C** $n - 6$

    D $n + 6$ **E** $n + 5$

25

y

K L

N

M (4, 6, 7)

O

P x

R Q

z

Diagram of cuboid **NOT** accurately drawn

The point M has the coordinates (4, 6, 7).

Work out the coordinates of the point Q.

    A (0, 6, 7)   **B** (4, 6, 0)   **C** (0, 0, 7)

    D (4, 0, 0)   **E** (4, 0, 7)

Check your answers on page 168. For full worked solutions see the CD.

Stage 2

Calculator

1 (a) Here is a prism made from cubes of 1 cm³

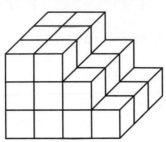

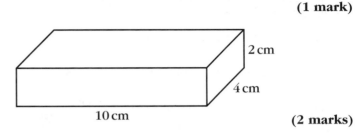

1 cm³

Find the volume of the prism. **(1 mark)**

(b) Here is a cuboid.
The length is 10 cm.
The width is 4 cm.
The height is 2 cm.
Work out the volume of this cuboid.

2 cm

4 cm

10 cm

(2 marks)

(Total 4 marks)

2 34 people were on a bus.

18 people got off.
14 people got on.

How many people are now on the bus? **(Total 2 marks)**

3 (a) Find the square of 11 **(1 mark)**

(b) Find $\sqrt{1.69}$ **(1 mark)**

(c) Work out $4 - 7$ **(1 mark)**

(d) Work out $-5 - -2$ **(1 mark)**

(Total 4 marks)

4 Each point on the graph represents the size
of a shoe and its length, in cm.

(a) Write down the length of
size 9 shoe. **(1 mark)**

(b) Write down the size of a shoe
with a length of 29 cm. **(1 mark)**

(c) Write down the length of a
size $8\frac{1}{2}$ shoe. **(1 mark)**

(Total 2 marks)

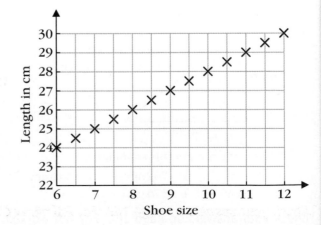

5 (a) Write 78.9 correct to one significant figure. (1 mark)

(b) Work out 5×3^2 (1 mark)

(Total 2 marks)

6 (a) Expand $3(2b + 5)$ (1 mark)

(b) Factorise $5g + 15$ (1 mark)

(c) Expand and simplify $(x + 2)(x + 5)$ (2 marks)

(Total 4 marks)

7 The length of a pen is measured as 13 cm to the nearest centimetre.

Write down the least length the pen could be. (Total 1 mark)

8 (a) Work out $3^4 \times 3^2$ (1 mark)

(b) $5x^2 = 80$ Find the value of x. (1 mark)

(Total 2 marks)

9

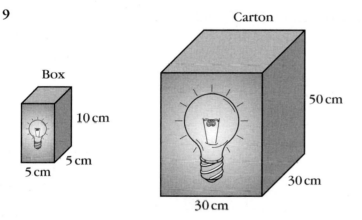

Carton

Box

Diagram **NOT** accurately drawn

50 cm

10 cm

5 cm

5 cm

30 cm

30 cm

A light bulb box measures 5 cm by 5 cm by 10 cm.
Light bulb boxes are packed into cartons.
A carton measures 30 cm by 30 cm by 50 cm.

Work out the number of light bulb boxes which can completely fill **one** carton. (Total 3 marks)

TOTAL FOR PAPER: 25 MARKS

END

Check your answers on page 168. For full worked solutions see the CD.

Working with fractions

- To **add** or **subtract** fractions, find **equivalent fractions** that have the same **denominator** (bottom number).

$$\overset{\times 3}{\overbrace{\frac{1}{2}}} + \frac{4}{6} = \underset{\times 3}{\underbrace{\frac{3}{6}}} + \frac{4}{6} = \frac{7}{6}$$

$$\frac{4}{6} - \overset{\times 3}{\overbrace{\frac{1}{2}}} = \frac{4}{6} - \underset{\times 3}{\underbrace{\frac{3}{6}}} = \frac{1}{6}$$

Key words

- equivalent fractions ☐
- denominator ☐
- numerator ☐
- mixed number ☐

- To add or subtract **mixed numbers**, deal with the whole number parts first.

Example

Work out　(a) $\frac{2}{3} + \frac{3}{5}$　**Grade E**　　　　(b) $5\frac{3}{4} - 2\frac{2}{3}$　**Grade C**

Write two lists of equivalent fractions.

⯆

Look for fractions with the same denominator.

⯆

Add the numerators.

⯆

Write as a mixed number.

(a) $\frac{2}{3} = \frac{4}{6} = \frac{6}{9} = \frac{8}{12} = \boxed{\frac{10}{15}} = \frac{12}{18}$

$\frac{3}{5} = \frac{6}{10} = \boxed{\frac{9}{15}} = \frac{12}{20}$

$\frac{2}{3} + \frac{3}{5} = \frac{10}{15} + \frac{9}{15} = \frac{19}{15}$

$= 1\frac{4}{15}$

First subtract the whole numbers.

⯆

Then subtract the fractions.

⯆

Now put the whole numbers and fractions back together.

(b) $5 - 2 = 3$

$\frac{3}{4} - \frac{2}{3} = \frac{9}{12} - \frac{8}{12} = \frac{1}{12}$

3 and $\frac{1}{12}$ is $3\frac{1}{12}$

So $5\frac{3}{4} - 2\frac{2}{3} = 3\frac{1}{12}$

TIP

$\frac{3}{4} = \frac{6}{8} = \boxed{\frac{9}{12}} = \frac{12}{16}$

$\frac{2}{3} = \frac{4}{6} = \frac{6}{9} = \boxed{\frac{8}{12}} = \frac{10}{15}$

- To **multiply** two fractions, multiply the numerators together and multiply the denominators together.

$$\frac{3}{4} \times \frac{4}{7} = \frac{12}{28}$$

- To **divide** fractions, invert the dividing fraction (turn it upside down) and multiply.

Turn ÷ into ×

$$\frac{1}{4} \div \frac{2}{5} = \frac{1}{4} \times \frac{5}{2} = \frac{5}{8}$$

Invert (turn upside down)

Key words

- improper fraction ☐
- mixed number ☐

- To multiply or divide **mixed numbers**, first change them to **improper fractions**.

$$1\frac{1}{2} \times 2\frac{1}{3} = \frac{3}{2} \times \frac{7}{3}$$

Example Work out (a) $2\frac{1}{4} \times 1\frac{1}{5}$ (b) $5\frac{1}{2} \div 1\frac{5}{6}$

Write as improper fractions. ⟶ (a) $2\frac{1}{4} = \frac{9}{4}$ and $1\frac{1}{5} = \frac{6}{5}$

▼

Multiply the top numbers and the bottom numbers. ⟶ $\frac{9}{4} \times \frac{6}{5} = \frac{54}{20}$

▼

Change to a mixed number then simplify the fraction part. ⟶ $= 2\frac{14}{20} = 2\frac{7}{10}$

Write as improper fractions. ⟶ (b) $5\frac{1}{2} = \frac{11}{2}$ and $1\frac{5}{6} = \frac{11}{6}$

▼

Invert the dividing fraction and change ÷ to ×. ⟶ $\frac{11}{2} \div \frac{11}{6} = \frac{11}{2} \times \frac{6}{11}$

TIP
11 and 2 are common factors. You could cancel *before* multiplying.
$\frac{\overset{1}{11}}{2} \times \frac{6}{\underset{1}{11}}$
$= \frac{\overset{3}{6}}{1\underset{1}{2}} = 3$

▼

Multiply the top numbers and the bottom numbers. Simplify the fraction. ⟶ $= \frac{66}{22} = 3$

Practice

1 Work out $\frac{3}{5}$ of £40

4 Work out

(a) $\frac{3}{4} \times \frac{7}{12}$ (b) $3\frac{3}{4} \times 1\frac{1}{3}$

2 Work out

(a) $\frac{1}{3} + \frac{3}{4}$

5 Work out

(a) $\frac{5}{12} \div \frac{3}{10}$ (b) $3\frac{1}{2} \div 5\frac{1}{4}$

(b) $2\frac{3}{8} + 5\frac{5}{6}$

6

A ——— $2\frac{3}{4}$ miles ——— B ——— C

3 Work out

(a) $\frac{7}{8} - \frac{1}{3}$

The diagram shows three towns, A, B and C.
The distance from town A to town C is $4\frac{1}{3}$ miles.
How far is it from town B to town C?

(b) $5\frac{2}{3} - 2\frac{1}{4}$

Check your answers on page 168. For full worked solutions see the CD.
See the Student Book on the CD if you need more help.

Question	1	2a	2b	3a	3b	4	5	6
Grade	F	E	C	D	C	C	C	C
Student Book pages	U3 18–19	U3 21–25		U3 21–25		U3 27–28	U3 27–28	U3 21–25

Percentages, fractions and decimals

- To compare **fractions**, **decimals** and **percentages** you can change them all to percentages.

- To write a decimal as a percentage you multiply by 100.

- To write a fraction as a percentage you change it to a decimal first by dividing the **denominator** into the **numerator**.

Key words

percentage ☐		numerator ☐	
decimal ☐		denominator ☐	
fraction ☐			

Example   Write these numbers in order of size, smallest first: **Grade E**

   0.77,   72%, $\frac{3}{4}$, $\frac{4}{5}$,   79%

Change all the numbers to percentages.

$0.77 \times 100 = 77$ so $0.77 = $ **77%**

72% is already a percentage.
$\frac{3}{4} = 3 \div 4 = 0.75 = $ **75%**
$\frac{4}{5} = 4 \div 5 = 0.8 = $ **80%**
79% is already a percentage.

TIP
Multiply by 100 to change a decimal to a percentage.

TIP
Divide the bottom number into the top number to change a fraction to a decimal.

EXAMINER'S TIP
Always show your working so that you can gain marks for the correct method.

Write the percentages in order. → 72%, 75%, 77%, 79%, 80%

Write the original numbers in the correct order. → 72%, $\frac{3}{4}$, 0.77, 79%, $\frac{4}{5}$

WATCH OUT!
Don't forget this final step.

- To compare different **proportions** you can change them all to percentages, so you are comparing like with like.

Key word

proportion ☐

Example   Jack scored 50 out of 75 in science, 56 out of 80 in maths and   **Grade D**
45 out of 60 in English.
In which subject did he do best?

Change all the marks into fractions and then into decimals by dividing.
Change the decimals into percentages by multiplying by 100.
Compare the percentages.

Science: $\frac{50}{75} = 50 \div 75 = 0.66\ldots = $ **66.7%**

Maths: $\frac{56}{80} = 56 \div 80 = 0.7 = $ **70%**

English: $\frac{45}{60} = 45 \div 60 = 0.75 = $ **75%**

He did best in English.

- To find a percentage of an amount you can:
  - change the percentage to a fraction and multiply *or*
  - change the percentage to a decimal and multiply *or*
  - work from 10%

Example
Work out 35% of £140

Grade E

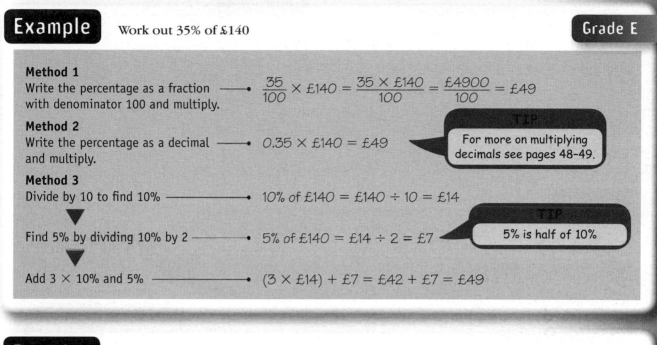

Method 1
Write the percentage as a fraction with denominator 100 and multiply. ⟶ $\frac{35}{100} \times £140 = \frac{35 \times £140}{100} = \frac{£4900}{100} = £49$

Method 2
Write the percentage as a decimal and multiply. ⟶ $0.35 \times £140 = £49$

TIP
For more on multiplying decimals see pages 48–49.

Method 3
Divide by 10 to find 10% ⟶ 10% of £140 = £140 ÷ 10 = £14

Find 5% by dividing 10% by 2 ⟶ 5% of £140 = £14 ÷ 2 = £7

TIP
5% is half of 10%

Add 3 × 10% and 5% ⟶ (3 × £14) + £7 = £42 + £7 = £49

Practice

1 Write these numbers in order of size, smallest first:

67%, $\frac{2}{3}$, $\frac{3}{5}$, 0.65, 63%

Grade E

2 Work out 15% of **(a)** 60 kg   **(b)** £160

Grade E

3 Work out 60% of £80

Grade E

4 Work out $7\frac{1}{2}$% of £200

Grade E

5 Bobbi scored 45 out of 60 in French, 60 out of 90 in German and 60 out of 80 in Spanish.
In which language did she do best?

Grade D

Check your answers on page 168. For full worked solutions see the CD.
See the Student Book on the CD if you need more help.

Question	1	2	3	4	5
Grade	E	E	E	E	D
Student Book pages	U3 32–34	U3 34–37	U3 34–37	U3 34–37	U3 32–34

Using percentages

- To **increase** a number by a percentage, you find the percentage of that number and then add this to the starting number.

- To **decrease** a number by a percentage, you find the percentage of that number and then subtract this from the starting number.

- **Sale prices** and **discounts** involve percentage decreases.

- **VAT** and **interest** involve percentage increases.

Key words

percentage increase and decrease ☐	VAT ☐
discount ☐	interest rate ☐
sale price ☐	percentage change ☐

Example

(a) A shop gives 20% discount on electrical goods. Work out the price of a TV that normally costs £80.

(b) Jay has to pay a bill of £80 plus VAT. What is the total bill?

Grade D

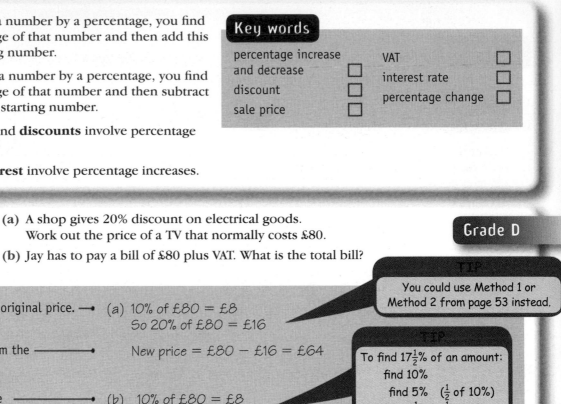

Find 20% of the original price. ——► (a) 10% of £80 = £8
So 20% of £80 = £16

TIP

You could use Method 1 or Method 2 from page 53 instead.

Subtract this from the original price. ——● New price = £80 − £16 = £64

TIP

To find $17\frac{1}{2}$% of an amount:
find 10%
find 5% ($\frac{1}{2}$ of 10%)
find $2\frac{1}{2}$% ($\frac{1}{2}$ of 5%)
Total = $17\frac{1}{2}$%

Find $17\frac{1}{2}$% of the original bill. ——● (b) 10% of £80 = £8
5% of £80 = £4
$2\frac{1}{2}$% of £80 = £2
$17\frac{1}{2}$% of £80 = £14

EXAMINER'S TIP

Make sure you understand percentages. There are always percentage calculations on GCSE papers.

Add this to the original amount. Total bill = £80 + £14
= £94

- To write one number as a percentage of another:
  1 write the amounts as a fraction   3 change the decimal to a percentage by multiplying by 100
  2 convert the fraction to a decimal

Example

Alice buys a watch for £40 and sells it for £50. What is her percentage profit?

Grade D

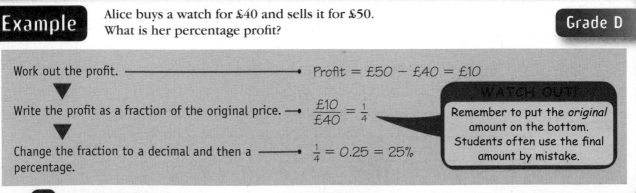

Work out the profit. ———————● Profit = £50 − £40 = £10

Write the profit as a fraction of the original price. ——► $\frac{£10}{£40} = \frac{1}{4}$

WATCH OUT!

Remember to put the *original* amount on the bottom. Students often use the final amount by mistake.

Change the fraction to a decimal and then a percentage. ———● $\frac{1}{4} = 0.25 = 25\%$

- An **index number** shows how a quantity changes over time.

- A **price index** shows how the price of something changes over time.
  - The index always starts at 100.
  - An index greater than 100 shows a price rise.
  - An index less than 100 shows a price fall.

Example

Grade C

The table shows the index numbers for the prices of houses in London over a 50-year period.

1960	1970	1980	1990	2000	2010
100	220	300	250	300	350

(a) A house cost £50 000 in 1960. How much would it have cost in 2000?

(b) What can you say about house prices between 1960 and 2000?

Divide the 1960 price by 100 to find 1%.
Then multiply by the index number for the year 2000.

(a) $\dfrac{£50\,000}{100} \times 300 = £150\,000$

(b) Prices went up until 1980, fell between 1980 and 1990, and then rose again after 1990.

Practice

Grade D

1   A shop reduces all its prices by 15%. Find the new cost of

    (a) a TV that normally costs £90

    (b) a DAB radio that normally costs £75

Grade D

2   Tom has to pay a garage bill of £160. VAT is added to the bill at $17\frac{1}{2}\%$.
    What is the total garage bill?

3   Jade invests £400 at 5% simple interest.

    (a) How much interest will Jade receive after 1 year?

    (b) How much interest will Jade receive after 3 years?

Grade D

4   A coat is reduced in price from £60 to £48.
    What is the percentage reduction?

Grade D

Check your answers on page 168 For full worked solutions see the CD.
See the Student Book on the CD if you need more help.

Question	1	2	3	4
Grade	D	D	D	D
Student Book pages	U3 34–37	U3 34–40	U3 37–40	U3 40–42

Fractions and percentages: topic test

Check how well you know this topic by answering these questions.
First cover the answers on the facing page.

Test questions

1 Find the difference between $\frac{3}{4}$ of 32 and $\frac{2}{3}$ of 27.

2 Rearrange these numbers in order of size, starting with the smallest:

$\frac{37}{100}$,   34%, $\frac{7}{20}$,   0.36

3 Henlow to Hitchin is $6\frac{1}{2}$ miles.
Clifton to Hitchin is $7\frac{1}{3}$ miles.
How much further is it to Hitchin from Clifton than from Henlow?

4 A television was reduced from £320 to £280 in a sale.
By what fraction was the price reduced in the sale?

5 Sharon bought a new car for £10 000.
It lost 30% of its value in 3 years.

   (a) Work out the loss in value.

   (b) What is the value of the car after 3 years?

6 Find the percentage reduction of the sunglasses.

SUNGLASSES

Sale Price **£8**

Normal Price £10

7 Jerry invested £200 at 8% simple interest for 2 years.
How much interest did she receive?

8 Narinda earns £250 a week.
She gets a pay rise of 4%.
How much will she now earn?

9 Work out the cost of a hot tub that costs £4000 plus VAT at 17.5%.

10 A trombone costs £450 plus VAT at $17\frac{1}{2}$%.
Work out the total cost of the trombone.

11 There are 56 kg of potatoes in a sack.
How many $2\frac{1}{2}$ kg bags can be filled from one sack?

12 Work out

  (a) $\frac{4}{9} + \frac{2}{9}$ **(b)** $\frac{7}{12} + \frac{1}{3}$ **(c)** $2\frac{1}{2} + 3\frac{1}{4}$

  (d) $\frac{7}{10} - \frac{3}{10}$ **(e)** $\frac{5}{6} - \frac{2}{5}$ **(f)** $4\frac{2}{3} - 2\frac{5}{8}$

  (g) $\frac{1}{3} \times \frac{1}{4}$ **(h)** $\frac{4}{5} \times \frac{3}{8}$ **(i)** $2\frac{1}{2} \times 3\frac{1}{4}$

  (j) $\frac{5}{9} \div \frac{3}{5}$ **(k)** $1\frac{1}{4} \div 3\frac{1}{2}$ **(l)** $15 \div \frac{3}{5}$

13 The table shows the index numbers for the average price of new small cars.

1960	1970	1980	1990	2000	2005
100	190	280	650	850	800

  (a) In which period was there the biggest increase in prices?

  (b) A new car cost £700 in 1960.
     What would a similar car have cost in 2000?

Now check your answers – see the facing page.

Cover this page while you answer the test questions opposite.

Worked answers

Revise this on...

E   **1** $\frac{3}{4}$ of 32 = 24 and $\frac{2}{3}$ of 27 = 18   24 − 18 = 6   page 97

E   **2**   Change into percentages:
$\frac{37}{100}$ = 37%,   34%, $\frac{7}{20}$ = 0.35 = 35%,   0.36 = 36%   page 98
The order is 34%, $\frac{7}{20}$, 0.36, $\frac{37}{100}$

D   **3** $7\frac{1}{3} - 6\frac{1}{2} = 7\frac{2}{6} - 6\frac{3}{6} = \frac{5}{6}$ It is $\frac{5}{6}$ mile further.   page 96

D   **4**   Reduction = £320 − £280 = £40   Fraction = $\frac{£40}{£320} = \frac{1}{8}$ page 100

D   **5**   (a)   Loss in value = 30% of £10 000   10% = £1000 so 30% = £3000   page 100
     (b)   £10 000 − £3000 = £7000

D   **6**   Reduction = £10 − £8 = £2   Percentage reduction = $\frac{£2}{£10} \times 100$ = 20%   page 100

D   **7**   Interest for 1 year = 8% of £200   10% of £200 = £20   page 100
                               2% of £200 = £20 ÷ 5 = £4
                                 8% = £20 − £4 = £16
     Interest for 2 years = 2 × £16 = £32

D   **8**   Increase in pay = 4% of £250 = $\frac{4}{100} \times £250 = \frac{£1000}{100}$ = £10   page 100
     Narindar will now earn £250 + £10 = £260

D   **9**   VAT = $17\frac{1}{2}$% of £4000   **10**   VAT = 17½% of £450   page 100
         10% of £4000 = £400   10% of £450 = £45
         5% of £4000 = £200   5% of £450 = £22.50
         $2\frac{1}{2}$% of £4000 = £100 $2\frac{1}{2}$% of £450 = £11.25
     So $17\frac{1}{2}$% of £4000 = £700   So $17\frac{1}{2}$% of £450 = £78.75
     Total = £4000 + £700 = £4700   Total = £450 + £78.75 = £528.75

C   **11** $56 \div 2\frac{1}{2} = \frac{56}{1} \div \frac{5}{2} = \frac{56}{1} \times \frac{2}{5} = \frac{112}{5} = 22\frac{2}{5}$ pages 96
     22 full sacks of potatoes, with 1 kg left over

E   **12**   (a) $\frac{4}{9} + \frac{2}{9} = \frac{6}{9}$ (b) $\frac{7}{12} + \frac{4}{12} = \frac{11}{12}$ (c) $5\frac{2+1}{4} = 5\frac{3}{4}$ (d) $\frac{7}{10} - \frac{3}{10} = \frac{4}{10} = \frac{2}{5}$ pages 96

D   (e) $\frac{25-12}{30} = \frac{13}{30}$ (f) $2\frac{16-15}{24} = 2\frac{1}{24}$ (g) $\frac{1 \times 1}{3 \times 4} = \frac{1}{12}$ (h) $\frac{4 \times 3}{5 \times 8} = \frac{12}{40} = \frac{3}{10}$

C   (i) $\frac{5}{2} \times \frac{13}{4} = \frac{65}{8} = 8\frac{1}{8}$ (j) $\frac{5}{9} \times \frac{5}{3} = \frac{25}{27}$

C   (k) $\frac{5}{4} \div \frac{7}{2} = \frac{5}{4} \times \frac{2}{7} = \frac{10}{28} = \frac{5}{14}$ (l) $\frac{15}{1} \div \frac{3}{5} = \frac{15}{1} \times \frac{5}{3} = \frac{75}{3} = 25$

C   **13**   (a) 1980–1990   (b) $\frac{£700}{100} \times 850 = £5950$ page 101

Tick the questions you got right.

Question	1	2	3	4	5	6	7	8	9	10	11	12a–d	12e–h	12i–l	13
Grade	E	E	D	D	D	D	D	D	D	D	C	E	D	C	C

Mark the grade you are working at on your revision planner on page xi.

Ratio

- A **ratio** is a way of comparing two numbers or quantities.

- To **simplify** a ratio you divide both its numbers by a **common factor**.

- When a ratio cannot be simplified it is in its **lowest terms**.

- Two ratios are **equivalent** when they both simplify to the same ratio.

Key words

ratio	☐
simplify	☐
common factor	☐
lowest terms	☐
equivalent ratios	☐

Example   Simplify these ratios.   **(a)** $18:12$ **(b)** $15:10$

Grade E

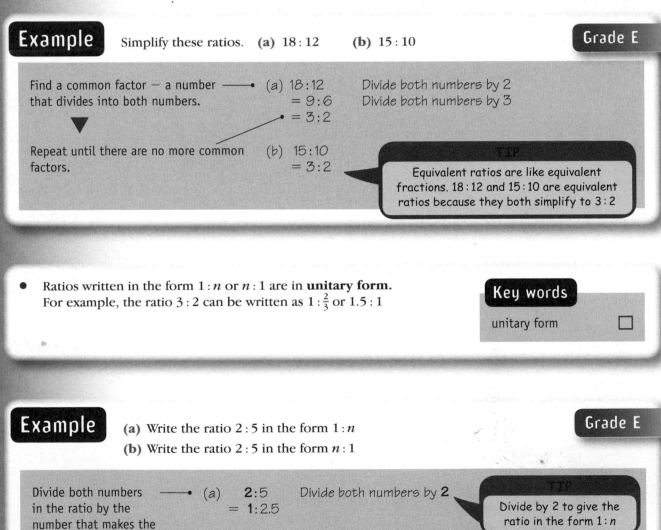

Find a common factor — a number that divides into both numbers.   → (a) $18:12$ Divide both numbers by 2
$$= 9:6 \quad \text{Divide both numbers by 3}$$
$$= 3:2$$

Repeat until there are no more common factors.   (b) $15:10$
$$= 3:2$$

TIP
Equivalent ratios are like equivalent fractions. $18:12$ and $15:10$ are equivalent ratios because they both simplify to $3:2$

- Ratios written in the form $1:n$ or $n:1$ are in **unitary form**.
  For example, the ratio $3:2$ can be written as $1:\frac{2}{3}$ or $1.5:1$

Key words

unitary form	☐

Example   **(a)** Write the ratio $2:5$ in the form $1:n$
           (b) Write the ratio $2:5$ in the form $n:1$

Grade E

Divide both numbers in the ratio by the number that makes the correct number into 1.   → (a)   **2:5**   Divide both numbers by **2**
$$= 1:2.5$$

(b)   **2:5**   Divide both numbers by **5**
$$= 0.4:1$$

TIP
Divide by 2 to give the ratio in the form $1:n$

TIP
Divide by 5 to give the ratio in the form $n:1$

- To share an amount in a given ratio:
  - find the total of all the numbers in the ratio
  - split the amount into fractions with that total as denominator.

Grade E

Grade C

Example

Deepal and Colette share £35 in the ratio 3 : 2

(a) What fraction of the amount does each receive?

(b) How much money does each receive?

Add the two numbers in the ratio to give a total. Write each share as a fraction with this total as denominator.	(a) $3 + 2 = 5$ Deepal receives $\frac{3}{5}$ Colette receives $\frac{2}{5}$

TIP
This means that the £35 is divided into **fifths**.

Divide the amount to be shared by this total. This is one part. Multiply this by the number of parts each person receives.	(b) One-fifth of £35 = £35 ÷ **5** = £7 Deepal receives $\frac{3}{5}$ = **3** × £7 = £21 Colette receives $\frac{2}{5}$ = **2** × £7 = £14

Practice

1   Write these ratios in their lowest terms.

    (a) 6 : 4   (b) 10 : 5   (c) 12 : 8   (d) 20 : 15

Grade D

2   Which of the ratios in question 1 are equivalent ratios?

Grade E

3   (a) Write these ratios in the form $n : 1$ (i) 5 : 4   (ii) 2 : 3
    (b) Write these ratios in the form $1 : n$ (i) 5 : 4   (ii) 2 : 3

Grade E

4   Wayne and Tracey share £24 in the ratio 5 : 3

    (a) What fraction of £24 does Wayne receive?

Grade E

    (b) How much money does each receive?

Grade C

5   Mick and Sophie shared a legacy in the ratio 3 : 2.
    Mick received £1500.
    How much did Sophie receive?

Grade C

Check your answers on page 168. For full worked solutions see the CD.
See the Student Book on the CD if you need more help.

Question	1	2	3	4a	4b	5
Grade	D	E	E	E	C	C
Student Book pages	U3 42–43	U3 42–44	U3 44–46	U3 50–51		U3 43–44

Proportion

- Two quantities are in **direct proportion** if their **ratio** stays the same when the quantities increase or decrease.

- In the **unitary method**, you find the value of *one* item first.

Key words

ratio ☐
proportion ☐
unitary method ☐

Example — Seven pens cost 84p.
How much do ten identical pens cost?

Grade D

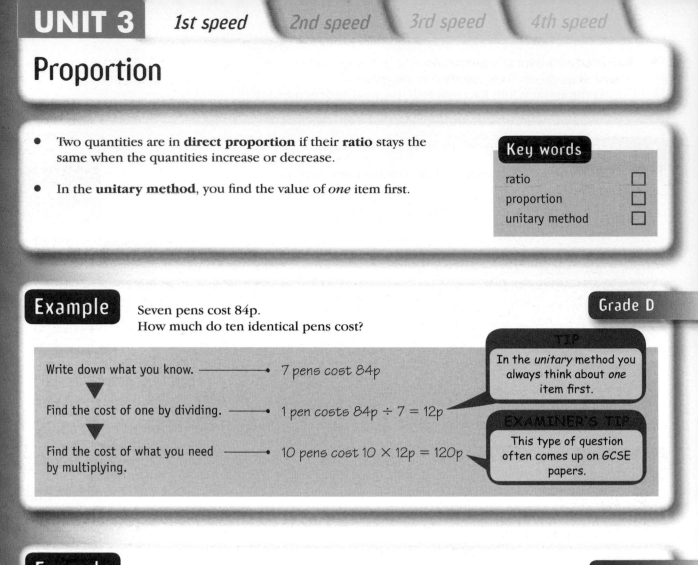

Write down what you know. ——→ 7 pens cost 84p

TIP
In the *unitary* method you always think about *one* item first.

Find the cost of one by dividing. ——→ 1 pen costs 84p ÷ 7 = 12p

EXAMINER'S TIP
This type of question often comes up on GCSE papers.

Find the cost of what you need ——→ 10 pens cost 10 × 12p = 120p
by multiplying.

Example — Here is a list of ingredients for making 12 cakes:

Grade D

200 g sugar
200 g butter
800 g flour
2 eggs
100 g dried fruit

(a) How many eggs would you need to make 18 cakes?

(b) How many grams of sugar would you need to make 9 cakes?

Look at the ratio of the ——→ (a) 12 : 18
numbers of cakes.

× 1½

12 cakes take 2 eggs
18 cakes take 2 × 1½ = 3 eggs

Use the unitary method. ——→ (b) 12 cakes take 200 grams

1 cake takes $\frac{200}{12}$ grams

9 cakes take 9 × $\frac{200}{12}$ = 3 × $\frac{200}{4}$

= 3 × 50

= 150 grams

EXAMINER'S TIP
Changing a recipe for a different number of people is often tested in GCSE papers.

- Ratios called **scales** are used to show the relationship between distances on a map and distances on the ground.

Example

Grade D

Two towns are 8.5 cm apart on a map.
The scale of the map is 1 : 50 000
How far apart are the towns in real life?

Write down the scale of the map. ——→ 1 cm represents 50 000 cm

Multiply by the map ——→ 8.5 cm represents 8.5 × 50 000 cm
distance.   = 425 000 cm

TIP
Divide by 100 to change cm to metres.

Change to suitable units. ——→ 425 000 cm = 4250 m
4250 m = 4.25 km

TIP
Divide by 1000 to change metres to km.

Practice

1   Three packets of sweets cost £2.40
    Work out the cost of five packets of sweets.

    Grade D

2   Here is a list of ingredients for making 10 cakes:

    Grade D

        200 g sugar
        200 g butter
        800 g flour
        2 eggs

    (a) How many eggs would you need to make 25 cakes?

    (b) How many grams of sugar would you need to make 15 cakes?

3   Two villages are 4 cm apart on a map.
    The scale of the map is 1 : 25 000
    How far apart are the two villages in real life?

    Grade D

Check your answers on page 168. For full worked solutions see the CD.
See the Student Book on the CD if you need more help.

Question	1	2	3
Grade	D	D	D
Student Book pages	U3 46–47	U3 48–50	UE 43–44

Ratio and proportion: topic test

Check how well you know this topic by answering these questions.
First cover the answers on the facing page.

Test questions

1 When ten people hire a boat for a weekend the cost per person is £56.
Work out the cost per person when the same boat is hired for a weekend by

(a) 5 people

(b) 7 people

(c) 8 people.

2 Kate is paid £100.80 for 16 hours' work in a café.
How much should she be paid for 12 hours' work?

3 The ingredients for making 15 cakes are:

250 g flour   175 g butter   175 g sugar

200 g fruit   3 eggs   330 ml milk

Jenny wants to make 10 of these cakes. Change the amounts given in the recipe to those needed for 10 cakes.
Give all your answers to the nearest whole unit.

4 Two towns are 7.5 cm apart on a map.
The scale of the map is 1 : 50 000
How far apart are the towns in real life?

5 Fatima has 35 CDs and DVDs.
The ratio of the number of CDs to the number of DVDs is 3 : 4
Work out how many CDs she has.

6 Given that 5 miles is equivalent to 8 kilometres

(a) work out, in kilometres
   (i) 40 miles   (ii) 12 miles   (iii) 240 miles

(b) work out, in miles
   (i) 24 kilometres   (ii) 50 kilometres   (iii) 3 kilometres

7 The depths of two wells are in the ratio 7 : 9
The depth of the deeper of the two wells is 63 metres.
Work out the depth of the other well.

Now check your answers – see the facing page.

Cover this page while you answer the test questions opposite.

Worked answers

Revise this on...

D 1 $10 \times £56 = £560$ for the hire of the boat — page 106
    (a) $£560 \div 5 = £112$ each for 5 people
    (b) $£560 \div 7 = £80$ each for 7 people
    (c) $£560 \div 8 = £70$ each for 8 people

D 2 16 hours for £100.80 — page 106
    1 hour for $£100.80 \div 16 = £6.30$
    12 hours for $£6.30 \times 12 = £75.60$
    She should be paid £75.60 for 12 hours' work.

D 3 Ratio $15:10 = 3:2$ ➔ Divide by 3 and multiply by 2 — page 106
    Flour $250\,g \div 3 \times 2 = 167\,g$ Butter $175 \div 3 \times 2 = 117\,g$
    Sugar $175 \div 3 \times 2 = 117\,g$ Fruit $200 \div 3 \times 2 = 133\,g$
    Eggs $3 \div 3 \times 2 = 2$ eggs   Milk $330 \div 3 \times 2 = 220\,ml$

D 4 $7.5\,cm \times 50\,000 = 375\,000\,cm$ — page 107
                        $= 3750\,m$ Divide by 100
                          $= 3.75\,km$ Divide by 1000

C 5 $3 + 4 = 7$ parts ➔ $35 \div 7 = 5$ ➔ $3 \times 5 = 15$ — page 105
    She has 15 CDs.

C 6 **(a)** 5 miles $= 8\,km$ ➔ 1 mile $= 8\,km \div 5 = 1.6\,km$ — page 106
        (i) $40 \times 1.6 = 64\,km$ **(ii)** $12 \times 1.6 = 19.2\,km$
        (iii) $240 \times 1.6 = 384\,km$
    (b) $8\,km = 5$ miles ➔ $1\,km = 5 \div 8 = 0.625$ miles
        (i) $24 \times 0.625 = 15$ miles   **(ii)** $50 \times 0.625 = 31.25$ miles
        (iii) $3 \times 0.625 = 1.875$ miles

C 7 Ratio of shallow to deep is $7:9$ — page 106
    9 'parts' $= 63\,m$ ➔ 1 'part' $= 7\,m$ ➔ 7 'parts' $= 7 \times 7 = 49\,m$
    The shallower well is 49 m deep.

Tick the questions you got right.

Question	1	2	3	4	5	6	7
Grade	D	D	D	D	C	C	C

Mark the grade you are working at on your revision planner on page xi.

Number: subject test

Check how well you know this topic by answering these questions.

Exam practice questions

1   Snack bars cost 45p each.
Donna spends £20 on snack bars.
What is the highest number of snack bars she can buy?

2   Write these numbers in order of size, smallest first.

     0.81, 82%, $\frac{2}{3}$, 65%, $\frac{4}{5}$

3   Jamie invests £400 at 5% simple interest for 3 years.
How much interest does Jamie receive?

4   Lewis wants to buy a new pair of trainers.
There are 3 shops that sell the trainers he wants.

Sports '4' all
Trainers
£5
plus 10 payments of
£4·50

EDEXCEL SPORTS
Trainers
$\frac{1}{5}$ **off**
usual price of
£65

Keef's Sports
Trainers
£50
plus
VAT at $17\frac{1}{2}\%$

(a)   Work out the cost of a pair of the trainers in Sports '4' all.

(b)   Work out the cost of a pair of the trainers in Edexcel Sports.

(c)   Work out the cost of a pair of the trainers in Keef's Sports.

5   (a)   Work out 35% of £40   (b)   Simon buys a TV for £64 plus VAT at $17\frac{1}{2}\%$
                                               Work out the total cost of the TV.

6   Rashmi buys 5 identical tins of paint for £35.
Work out the cost of 8 of these tins of paint.

7   Here is part of Ashanti's gas bill.
Work out how much Ashanti has to pay
for the units of gas he has used.

Gas bill	
New reading	54 516 units
Old reading	53 939 units
Price per unit	45p

8   Bob lays 200 bricks in 1 hour.
He always works at the same speed.

Work out how long it will take Bob to lay 960 bricks.
Give your answer in hours and minutes.

9   Tom makes concrete.
He mixes gravel and cement in the ratio 6 : 1

One day Tom used 3 bags of cement.

(a) How many bags of gravel did he need?

Another day Tom used 48 bags of gravel.

(b) How many bags of cement did he need?

10   Two towns are 6.7 cm apart on a map.
The scale of the map is 1 : 50 000
How far apart are the towns in real life?

11   (a) Work out $\frac{3}{4}$ of £36

(b) Write these fractions in order of size, smallest first:

$\frac{2}{3}, \frac{3}{5}, \frac{3}{4}, \frac{1}{2}$

(c) Work out $\frac{2}{3} - \frac{1}{4}$

(d) Work out $3\frac{3}{5} \times 2\frac{2}{9}$

12   Rachel shares £32 between her children Marco and José in the ratio 5 : 3
How much money does each child receive?

13   A large tub of popcorn costs £3.80 and holds 200 g.
A regular tub of popcorn costs £3.50 and holds 175 g.

Rob says that the 200 g large tub is the better value
for money
Linda says that the 175 g regular tub is the better
value for money.

Who is correct?
Explain the reasons for your answer.
You must show all your working.

200g
Large

175g
Regular

£3.80   £3.50

Check your answers on page 168. For full worked solutions see the CD.

Tick the questions you got right.

Question	1	2	3	4a	4b	4c	5a	5b	6	7	8	9	10	11a	11b	11c	11d	12	13
Grade	F	E	E	F	E	D	E	D	D	D	D	D	D	F	E	D	C	C	C
Revise this on page	—	98	100	—	97	100	99	100	106	—	106	106	107	99	96	96	97	105	106

Mark the grade you are working at on your revision planner on page xi.

Go to the pages shown to revise for the ones you got wrong.

Algebra

- The ^{2} in 7^2 is called an **index** or **power**. It tells you how many times the given number must be multiplied by itself.
- The plural of index is **indices**.
- To **multiply** powers of the same number or letter, add the indices:
$$3^3 \times 3^4 = 3^{3+4} = 3^7 \qquad x^a \times x^b = x^{a+b}$$
- Any number or letter raised to the **power 1** is equal to the number or letter itself:
$$3^1 = 3 \qquad x^1 = x$$

Key words

power ☐
index ☐
indices ☐

Example   Simplify   (a) $p \times p \times p \times p$ **Grade D**   (b) $x^3 \times x^2$ **Grade C**

(c) $2y^2 \times 3y$ **Grade C**

Count the ps.
Write an index number to show how many times p is multiplied by itself.

(a) $p \times p \times p \times p = p^4$

TIP
$x^a \times x^b = x^{a+b}$

(b) **Method 1: Using the index laws**
Add the indices.

$x^3 \times x^2 = x^{3+2} = x^5$

Method 2: Writing out the multiplication
Write out the multiplication fully.

$\underbrace{x \times x \times x}_{x^3} \times \underbrace{x \times x}_{x^2}$

Count the xs and write the index number.

$= x^5$

TIP
y is y^1

(c) Multiply the numbers.

$2 \times 3 = 6$

Multiply the letter terms.

$y^2 \times y = y^{2+1} = y^3$

Combine the number and letter terms.

$2y^2 \times 3y = 6y^3$

- To **divide** powers of the same number or letter, subtract the indices:
$$4^5 \div 4^2 = 4^{5-2} = 4^3 \qquad x^a \div x^b = x^{a-b}$$

Example   Simplify $6y^6 \div 3y^2$ **Grade C**

Method 1: Using the index laws

$$6y^6 \div 3y^2$$

Divide the numbers.

$6 \div 3 = 2$

TIP
$x^a \div x^b = x^{a-b}$

Divide the letter terms.

$y^6 \div y^2 = y^{6-2} = y^4$

Combine the number and letter parts.

$6y^6 \div 3y^2 = 2y^4$

Method 2

Write the division like this. ⟶ $6y^6 \div 3y^2$

$$= \frac{6y^6}{3y^2} = \frac{6}{3} \times \frac{y \times y \times y \times y \times y \times y}{y \times y}$$

Look for common factors and cancel. ⟶ $= \frac{2}{1} \times \frac{\cancel{y} \times \cancel{y} \times y \times y \times y \times y}{\cancel{y} \times \cancel{y}}$

Write the answer as simply as possible. ⟶ $= 2 \times y \times y \times y \times y$
$= 2 \times y^4$
$= 2y^4$

TIP

6 and 3 have common factor 3, so cancel by 3. Two of the ys in the denominator cancel with two of the ys in the numerator.

TIP

$\frac{2}{1} = 2$

- To raise a power of a letter to a further power, multiply the indices: $(x^a)^b = x^{ab}$

Example   Simplify $(x^4)^2$ **Grade C**

Method 1: Using the index laws

Use $(x^a)^b = x^{ab}$ ⟶ $(x^4)^2 = x^{4 \times 2} = x^8$

Method 2: Writing out the multiplication

Write out the multiplication like this. ⟶ $(x^4)^2 = x^4 \times x^4$
Use $x^a \times x^b = x^{a+b}$ ⟶ $= x^8$

- Any non-zero number or letter raised to the **power 0** is equal to 1: $3^0 = 1$ $y^0 = 1$

Example   Simplify $2x^2 \div x^2$ **Grade C**

Divide the numbers. ⟶ $2 \div 1 = 2$

Divide the letter terms. ⟶ $x^2 \div x^2 = x^{2-2} = x^0 = 1$

Combine the number and letter parts. ⟶ $2 \times 1 = 2$

TIP

x^2 is the same as $1x^2$

Practice

Grade D   **1**   Simplify $t \times t \times t$

Grade C   **2**   Simplify

(a) $h^3 \times h^4$ (b) $\dfrac{12x^5}{3x^2}$

Grade C   **3**   Simplify

(a) $8x^5 \div 2x^3$ (b) $10y^4 \div 5y^2$

4   Simplify   **Grade C**

(a) $\dfrac{12q^6}{4q^2}$ (b) $\dfrac{20t^5}{5t^2}$

5   Simplify   **Grade C**

(a) $(a^3)^4$ (b) $9y^2 \div 3y$ (c) $(n^2)^2$

6   Simplify $4m^2n \times 2nm$ **Grade C**

Check your answers on page 168. For full worked solutions see the CD.
See the Student Book on the CD if you need more help.

Question	1	2	3	4	5	6
Grade	D	C	C	C	C	C
Student Book pages	U3 59–61	U3 59–61	U3 59–61	U3 59–61	U3 59–61	U3 59–61

Linear graphs

- A graph representing a **linear relationship** is always a straight line.

- **Distance–time graphs** are used to relate the distance travelled to the time taken, and to calculate speeds.

Key words

time ☐		distance–time graph ☐
distance ☐		average speed ☐

Example

The distance–time graph represents a cyclist's journey.

(a) At what time did the cyclist take a break?

(b) Why does the graph slope downwards after 6 pm?

(c) How far from home was the cyclist at 5 pm?

(d) What was the cyclist's average speed between 1 pm and 3 pm?

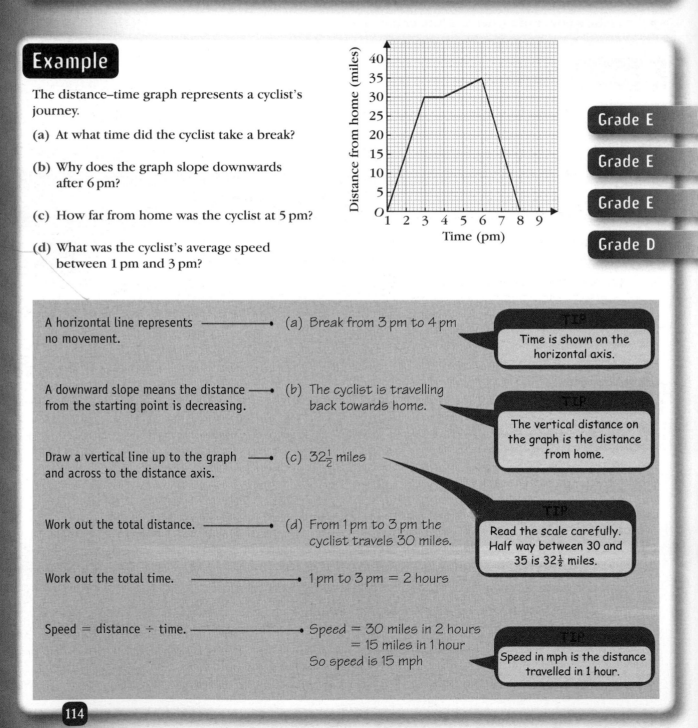

Grade E

Grade E

Grade E

Grade D

A horizontal line represents no movement. ⟶ (a) Break from 3 pm to 4 pm

> **TIP**
> Time is shown on the horizontal axis.

A downward slope means the distance from the starting point is decreasing. ⟶ (b) The cyclist is travelling back towards home.

> **TIP**
> The vertical distance on the graph is the distance from home.

Draw a vertical line up to the graph and across to the distance axis. ⟶ (c) $32\frac{1}{2}$ miles

Work out the total distance. ⟶ (d) From 1 pm to 3 pm the cyclist travels 30 miles.

> **TIP**
> Read the scale carefully. Half way between 30 and 35 is $32\frac{1}{2}$ miles.

Work out the total time. ⟶ 1 pm to 3 pm = 2 hours

Speed = distance ÷ time. ⟶ Speed = 30 miles in 2 hours
= 15 miles in 1 hour
So speed is 15 mph

> **TIP**
> Speed in mph is the distance travelled in 1 hour.

Practice

1 The graph represents a saleswoman's journey to and from a meeting.

    (a) How long did the meeting last?  Grade E

    (b) How far had the saleswoman travelled after 1 hour? Grade E

    (c) What was her average speed from $2\frac{1}{2}$ hours to 3 hours? Grade D

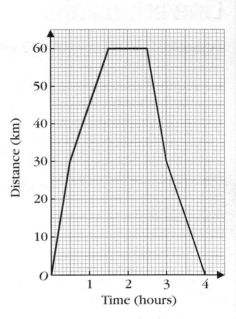

2 The graph represents the movement of a remote control car, to and from a starting point.

    (a) How far from the starting point did the car travel? Grade E

    (b) For how long was the car stationary? Grade E

    (c) What was the average speed of the car, in metres per second, during the first 10 seconds? Grade D

3 The graph shows part of Alison's journey to visit her grandmother.

    (a) What was Alison's speed between 13:00 and 14:00? Grade D

    (b) Alison travels home at an average speed of 40 km/h. Complete the travel graph. Grade C

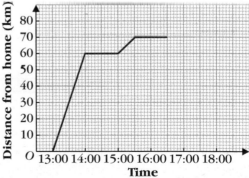

Check your answers on page 168. For full worked solutions see the CD.
See the Student Book on the CD if you need more help.

Question	1ab	1c	2ab	2c	3a	3b
Grade	E	D	E	D	D	C
Student Book pages	U3 93–96		U3 93–96		U3 93–96	

Curved graphs

- An equation containing an x^2-term is called a **quadratic equation**.

- The graph of a quadratic equation is a curved quadratic graph, or parabola.
  It has a symmetrical U- shape: ∪ or ∩.

Key words

| curve ☐ | quadratic ☐ |
| equation ☐ | coordinate ☐ |

Example   Grade C

(a) Complete the table of values for $y = 3x^2 - 3x - 2$.

x	-2	-1	0	1	2	3
y						

(b) On the grid, use your table of values to draw the graph of $y = 3x^2 - 3x - 2$.

(c) Write down the minimum value of y.

(d) Use your graph to find

   (i) the value of y when $x = 1.8$

   (ii) the values of x when $y = 6$

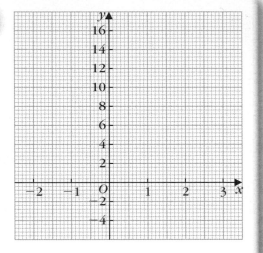

$y = 3x^2 - 3x - 2$ is a quadratic graph, a curve.
Work out each value of y. ●————

(a) When $x = \mathbf{3}$, $y = 3 \times \mathbf{3}^2 - 3 \times \mathbf{3} - 2 = 27 - 9 - 2 = 16$
When $x = \mathbf{2}$, $y = 3 \times \mathbf{2}^2 - 3 \times \mathbf{2} - 2 = 12 - 6 - 2 = 4$

TIP

Start working out the values from the right-hand (positive) side of the table. Try to spot symmetry in the numbers in the table: this will help you complete it.

x	-2	-1	0	1	2	3
y	16	4	-2	-2	4	16

This means the point $(-2, 16)$

Plot the table values on the grid.   **(b)**

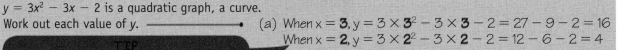

Join the points with a curve.

WATCH OUT!

The bottom must be curved. If there are two points at the bottom the curve should sink slightly below them. Students often wrongly draw a straight line.

The minimum value of y is at the bottom of the curve. Draw a line across to the y-axis.   **(c)** $y = -2.7$

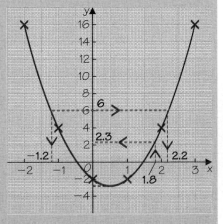

Draw a line from the x-value up to ●———
the line and across to the y-axis.

(d) (i) When $x = 1.8$,
       $y = 2.3$

Draw a line from the y-value across ●———
to the line and down to the x-axis.

(ii) When $y = 6$, $x = -1.2$
       and $x = 2.2$

Example

Water is poured into this container. Sketch a graph to show the relationship between the water level and the volume of water in the container.

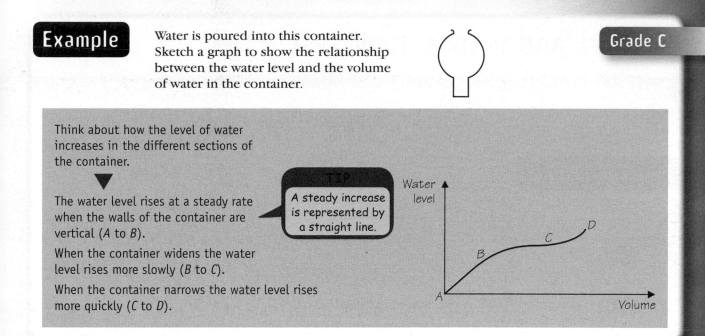

Think about how the level of water increases in the different sections of the container.

▼

The water level rises at a steady rate when the walls of the container are vertical (*A* to *B*).

TIP

A steady increase is represented by a straight line.

When the container widens the water level rises more slowly (*B* to *C*).

When the container narrows the water level rises more quickly (*C* to *D*).

Practice

Grade C

1 (a) Complete the table of values for $y = x^2 - 3x$.

x	-1	0	1	2	3	4
y						

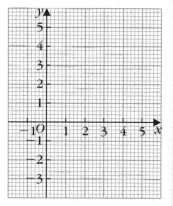

(b) Draw the graph of $y = x^2 - 3x$ on a copy of the grid.

(c) Use your graph to find

   (i) the value of y when $x = -\frac{1}{2}$

   (ii) the value of x when $y = 1.75$

Grade C

2 The sketch graph shows the sound level of a television from when it is switched on. Describe how the sound level changes between

(a) *A* and *B*

(b) *B* and *C*

(c) *C* and *D*.

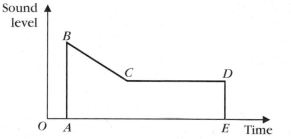

Check your answers on page 169. For full worked solutions see the CD.
See the Student Book on the CD if you need more help.

Question	1	2
Grade	C	C
Student Book pages	U3 100–105	U3 97–99

Algebra and graphs: topic test

Check how well you know this topic by answering these questions.
First cover the answers on the facing page.

Test questions

1   This graph shows a car's journey.

    (a) How far is the car from home after
        10 minutes?

    (b) For how long was the car stationary
        after 25 minutes?

    (c) What was the average speed, in km/h,
        from 15 minutes to 25 minutes?

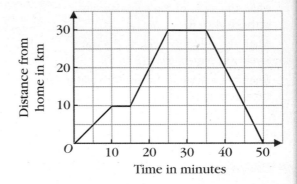

2   Simplify $p \times p \times p \times p$

3   (a) Complete the table of values for
       $y = x^2 - 3x - 1$.

x	-1	0	1	2	3	4
y						

    (b) On a copy of the grid, draw the graph of
       $y = x^2 - 3x - 1$.

    (c) Use your graph to find the values of x when
       $y = 0.6$

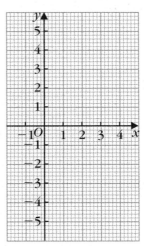

4   Simplify $d^3 \times d^5$

5   Simplify $3y^5 \div y^2$

6   Expand $2a(a + b)$

7   Expand $x(2x^2 + 1)$

Now check your answers – see the facing page.

Cover this page while you answer the test questions opposite.

Worked answers

Revise this on...

E   **1**   **(a)**   10 km   page 114

E   **(b)**   From 25 minutes the graph is horizontal for 2 divisions.   page 114
Each division represents 5 minutes so the car is stationary
for 10 minutes.

D   **(c)**   Between 15 minutes and 25 minutes the car travels 20 km.   page 114

$$\text{Speed} = 20 \text{ km in 10 minutes}$$
$$= 2 \text{ km per minute}$$
$$= 120 \text{ km per hour}$$

D   **2** $p \times p \times p \times p = p^4$ page 112

C   **3**   **(a)**   When $x = \mathbf{4}$, $y = \mathbf{4}^2 - 3 \times \mathbf{4} - 1 = 16 - 12 - 1 = 3$,   page 116
when $x = \mathbf{3}$, $y = \mathbf{3}^2 - 3 \times \mathbf{3} - 1 = 9 - 9 - 1 = -1$,
when $x = \mathbf{2}$, $y = \mathbf{2}^2 - 3 \times \mathbf{2} - 1 = 4 - 6 - 1 = -3$, ...

x	−1	0	1	2	3	4
y	3	−1	−3	−3	−1	3

(b)

$y = x^2 - 3x - 1$

(c)   From the graph, when $y = 0.6$, $x = -0.5$ and $x = 3.5$

C   **4** $d^3 \times d^5 = d^{3+5} = d^8$ page 112

C   **5** $3y^5 \div y^2 = 3y^{5-2} = 3y^3$ page 112

C   **6** $2a(a + b) = 2a \times a + 2a \times b = 2a^2 + 2ab$ pages 58, 112

C   **7** $x(2x^2 + 1) = x \times 2x^2 + x \times 1 = 2x^3 + x$ pages 58, 112

Tick the questions you got right.

Mark the grade you are working at on
your revision planner on page xi.

Question	1ab	1c	2	3	4	5	6	7
Grade	E	D	D	C	C	C	C	C

Formulae

- A **word formula** uses words to represent a relationship between quantities. For example:

    pay = rate of pay × hours worked

- An **algebraic formula** uses letters to represent a relationship between quantities. For example, the perimeter of a rectangle, P, is related to its length l and width w by

    $P = 2l + 2w$

Example

Grade F

The monthly charge for using a photocopier is given by the formula

$$C = 0.05 \times n + 80$$

where C is the total charge, in £, and n is the number of photocopies made. Find the total cost for January, when 2200 photocopies were made.

Identify the numbers to use in the formula. ———→ To find C, you need n = 2200

Write down the calculation. ———→ C = 0.05 × n + 80

Replace n with its number value. ———→ = 0.05 × 2200 + 80
= 110 + 80
= £190

Example

Grade D

The lengths of the sides of a triangle are shown in the diagram.

(a) Write down an expression in terms of x for the length of the perimeter.

(b) Write down a formula for the perimeter, P, in terms of x.

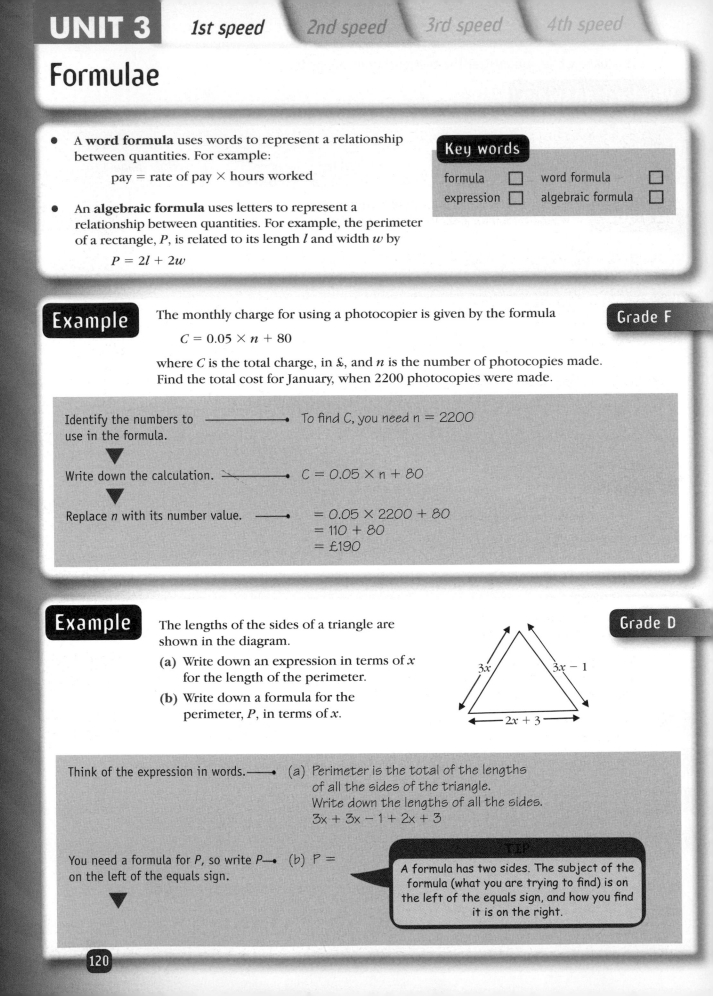

$3x$ $3x - 1$

$2x + 3$

Think of the expression in words. ———→ (a) Perimeter is the total of the lengths of all the sides of the triangle.
Write down the lengths of all the sides.
3x + 3x − 1 + 2x + 3

You need a formula for P, so write P ———→ (b) P =
on the left of the equals sign.

TIP
A formula has two sides. The subject of the formula (what you are trying to find) is on the left of the equals sign, and how you find it is on the right.

On the right of the equals sign write how you work it out (the sum of the side lengths).

$$P = 3x + 3x - 1 + 2x + 3$$

▼

Simplify by collecting like terms. ⟶• $= 8x + 2$

Grade E

Example

C is the total cost in pounds of buying n CDs at £12 each. Write down a formula for C in terms of n.

Think about the cost of different numbers of CDs. ⟶• Cost of 1 CD is £12
Cost of 2 CDs is 2 × £12

▼

What is the cost of n CDs? ⟶• Cost of n CDs is n × 12

▼

Write your formula using letters. ⟶• $C = n \times 12$

▼

Rewrite using correct algebraic notation (no × sign, numbers *before* letters). ⟶• $C = 12n$

TIP
Write the number first then the letter without the × sign.

Practice

Grade F

1   The total number of eggs in some boxes is given by the formula

total number of eggs = 12 × number of boxes

Find the number of eggs in 15 boxes.

Grade E

2   The charge £C for parking in an airport car park is given by the formula
$C = 2h + 3$
where h is the number of hours.
Calculate the charge for parking a car for 4 hours.

Grade E

3   The weight of one computer is 8 kg, and the weight of its packaging is k kg.
Altogether, n computers are sold. The total weight of computers and packaging is W.
Write down a formula for W in terms of k and n.

Grade C

4   The lengths of the sides of a quadrilateral are shown in the diagram.
Write down a formula for the perimeter, P, in terms of y.

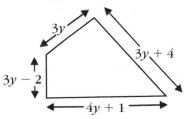

Check your answers on page 169. For full worked solutions see the CD.

See the Student Book on the CD if you need more help.

Question	1	2	3	4
Grade	F	E	E	C
Student Book pages	U3 112–114	U3 116–118	U3 116–118	U3 116–118

Rearranging formulae

- The **subject** of a **formula** appears on its own on one side of the formula and does not appear on the other side. For example:

$$t = 4l + 4 \text{ can be } \textbf{rearranged} \text{ to give } l = \frac{t - 4}{4}$$

t is the subject l is the subject

Key words

rearrange	☐
subject	☐
algebraic formula	☐

Example   Rearrange the formula $y = 6x + 2$ to make x the subject.   **Grade D**

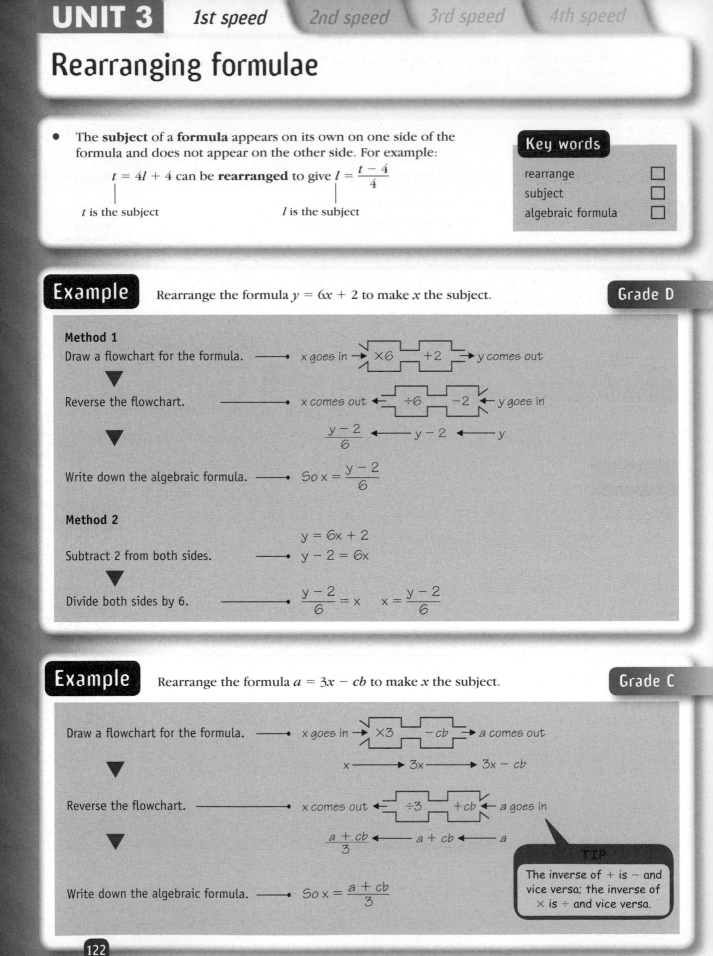

Method 1

Draw a flowchart for the formula. ——• x goes in → ×6 → +2 → y comes out

Reverse the flowchart. ——• x comes out ← ÷6 ← −2 ← y goes in

$\frac{y - 2}{6}$ ← $y - 2$ ← y

Write down the algebraic formula. ——• So $x = \dfrac{y - 2}{6}$

Method 2

$$y = 6x + 2$$

Subtract 2 from both sides. ——• $y - 2 = 6x$

Divide both sides by 6. ——• $\dfrac{y - 2}{6} = x \qquad x = \dfrac{y - 2}{6}$

Example   Rearrange the formula $a = 3x - cb$ to make x the subject.   **Grade C**

Draw a flowchart for the formula. ——• x goes in → ×3 → $- cb$ → a comes out

x → $3x$ → $3x - cb$

Reverse the flowchart. ——• x comes out ← ÷3 ← $+ cb$ ← a goes in

$\dfrac{a + cb}{3}$ ← $a + cb$ ← a

Write down the algebraic formula. ——• So $x = \dfrac{a + cb}{3}$

TIP

The inverse of + is − and vice versa; the inverse of × is ÷ and vice versa.

- You can **substitute** numbers into formulae. 'Substitute' means 'replace a letter with a number value.'

Example

Given the formula $P = bc + 8$
(a) find P when $b = -2$ and $c = 3$
(b) find b when $P = 20$ and $c = 4$.

Grade C

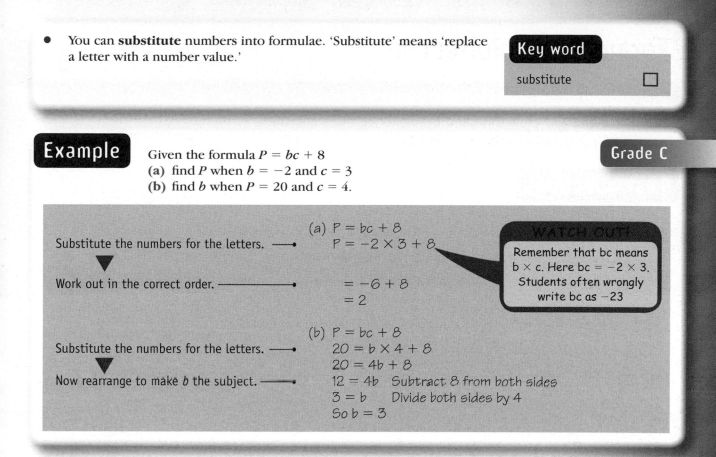

Substitute the numbers for the letters. ⟶ (a) $P = bc + 8$
$P = -2 \times 3 + 8$

▼

Work out in the correct order. ⟶ $= -6 + 8$
$= 2$

WATCH OUT!
Remember that bc means b × c. Here bc = -2 × 3. Students often wrongly write bc as -23

Substitute the numbers for the letters. ⟶ (b) $P = bc + 8$
$20 = b \times 4 + 8$
$20 = 4b + 8$

▼

Now rearrange to make b the subject. ⟶ $12 = 4b$ Subtract 8 from both sides
$3 = b$ Divide both sides by 4
So $b = 3$

Practice

1 (a) Using the formula $x = ab + c$, find the value of x when $a = 0.6$, $b = 10$ and $c = -2$.

   (b) Using the formula $y = 2c + d$, find the value of y when $c = 0.8$ and $d = -2$.

Grade D

2 (a) Rearrange $3x - 4 = y$ to make x the subject.
   (b) Rearrange $2x + 3 = y$ to make x the subject.

Grade D

3 (a) Rearrange $4x + bc = a$ to make x the subject.
   (b) Rearrange $2(x + 3) = t$ to make x the subject.

Grade C

4 Given the formula $v = u + at$, find the value of a when $v = 21.5$, $u = 4$ and $t = 7$.

Grade C

5 Rearrange the formula $y = \frac{x}{2} + 8$ to make x the subject.

Grade C

Check your answers on page 169. For full worked solutions see the CD.
See the Student Book on the CD if you need more help.

Question	1	2	3	4	5
Grade	D	D	C	C	C
Student Book pages	U3 116–118	U3 120–121	U3 120–121	U3 118–120	U3 120–121

Formulae: topic test

Check how well you know this topic by answering these questions.
First cover the answers on the facing page.

Test questions

1   The total cost of buying a number of pencils is given by the word formula

   total cost = number of pencils × cost of each pencil

   Find the total cost when the number of pencils is 15 and the cost of each pencil is 8p.

2   The total cost of hiring a piece of equipment is given by the formula

   total cost = number of hours × £15 + £30

   Find the total cost of hiring a piece of equipment for 8 hours.

3 $T = 2p + 3q$ Find T when $p = 4$ and $q = 5$.

4 C is the total cost of buying n books at £8 each.
   Write down a formula for the total cost of buying n books.

5 $H = 7c - 3d$ Find H when $c = 2$ and $d = -3$.

6 $y = 4x - 3$ Find x when $y = 7$.

7   Pens cost 20p each and pencils cost 15p each.
   T is the total cost, in pence, of buying r pens and q pencils.
   Write down a formula for T in terms of r and q.

8   Rearrange the formula $y = 5x - 20$ to make x the subject.

9   The diagram shows the lengths of the four sides
   of a quadrilateral, in terms of x.
   P is the length of the perimeter of the quadrilateral.
   Write down a formula for P in terms of x.

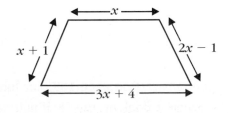

10   Rearrange the formula $t = \dfrac{v}{3} - 4$ to make v the subject.

Now check your answers – see the facing page.

Cover this page while you answer the test questions opposite.

Worked answers

G 1 Total cost = 15 × 8p = 120p — page 120

F 2 Total cost = 8 × £15 + £30 = £120 + £30 = £150 — page 120

E 3 T = 2 × 4 + 3 × 5 = 8 + 15 = 23 — page 123

E 4 C = n × 8   or   C = 8n — page 121

D 5 H = 7 × 2 − 3 × (−3) = 14 + 9 = 23 — page 123

D 6 y = 4x − 3 — page 123
When y = 7,   7 = 4x − 3
          4x = 7 + 3   Add 3 to both sides
          4x = 10
           x = 10 ÷ 4 = 2.5   Divide both sides by 4

D 7 The cost of r pens is r × 20 or 20r — page 123
The cost of q pencils is q × 15 or 15q
Total cost = T = 20r + 15q

D 8 y = 5x − 20 — page 122
5x = y + 20   Add 20 to both sides

$x = \dfrac{y + 20}{5}$ Divide both sides by 5

C 9 Perimeter = P = x + (x + 1) + (2x − 1) + (3x + 4) = 7x + 4 — page 120

C 10 $t = \dfrac{v}{3} - 4$ — page 122

$\dfrac{v}{3} = t + 4$ Add 4 to both sides

v = 3(t + 4)   Multiply both sides by 3

Tick the questions you got right.

Question	1	2	3	4	5	6	7	8	9	10
Grade	G	F	E	E	D	D	D	D	C	C

Mark the grade you are working at on your revision planner on page xi.

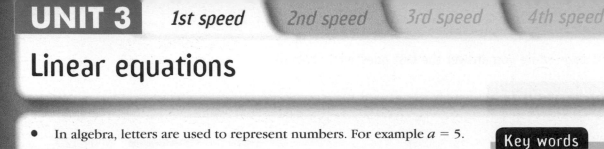

Linear equations

- In algebra, letters are used to represent numbers. For example $a = 5$.

- To **solve** an **equation** use the **balancing** method. You must do the same to each side.

$$a + 4 = 7 \quad \rightarrow \quad a + 4 - 4 = 7 - 4 \quad \rightarrow \quad a = 3$$
$$a - 3 = 1 \quad \rightarrow \quad a - 3 + 3 = 1 + 3 \quad \rightarrow \quad a = 4$$
$$5a = 30 \quad \rightarrow \quad 5a \div 5 = 30 \div 5 \quad \rightarrow \quad a = 6$$
$$\frac{a}{2} = 7 \quad \rightarrow \quad \frac{a}{2} \times 2 = 7 \times 2 \quad \rightarrow \quad a = 14$$

- In a combined equation, deal with the $+$ and $-$ first.

$$3a + 7 = 1 \rightarrow 3a + 7 - 7 = 1 - 7 \rightarrow 3a = -6 \rightarrow a = -2$$

Example   Solve $4x + 3 = 5$ Grade E

Use the balancing method to get
the letters on their own on one side.

$$4x + 3 = 5$$
$$4x + 3 - 3 = 5 - 3 \qquad \text{Take 3 from both sides}$$
$$4x = 2$$
$$x = 2 \div 4 \text{ or } \tfrac{2}{4} \qquad \text{Divide both sides by 4}$$

▼

Write the answer as simply
as possible.

$$x = \tfrac{1}{2} \text{ or } 0.5$$

WATCH OUT!

Make sure you divide the
right way around:
$2 \div 4$ not $4 \div 2$

Example   Solve $6 + 4x = 2x + 16$ Grade D

Use the balancing method to get
the letters on their own on one side.

$$6 + 4x = 2x + 16$$
$$6 + 4x - 2x = 2x - 2x + 16 \qquad \text{Take 2x from both sides}$$
$$6 + 2x = 16$$
$$6 - 6 + 2x = 16 - 6 \qquad \text{Take 6 from both sides}$$
$$2x = 10$$
$$x = 5 \qquad \text{Divide both sides by 2}$$

Example   Solve $\dfrac{3d}{4} + 5 = 50$ Grade C

Use the balancing method to get
the letters on their own on one side.

$$\frac{3d}{4} + 5 = 50$$
$$\frac{3d}{4} = 45 \qquad \text{Take 5 from both sides}$$
$$3d = 180 \qquad \text{Multiply both sides by 4}$$
$$d = 180 \div 3 = 60 \qquad \text{Divide both sides by 3}$$

- In an equation with **brackets**, **expand** the brackets first.

$$3(x + 1) = 4 \quad \rightarrow \quad 3x + 3 = 4$$

Key words

bracket ☐   expand ☐

Example

Solve $2(2x - 7) = 7$

Grade D

$$2(2x - 7) = 7$$
Expand the bracket first. ⟶ $2 \times 2x + 2 \times -7 = 7$
$$4x - 14 = 7$$

WATCH OUT!

Make sure you multiply *every* term inside the bracket. Students often wrongly multiply just the first term, giving $4x - 7$ instead of the correct expansion $4x - 14$.
For more on expanding brackets, see page 58.

Then use the balancing method. ⟶ $4x - 14 + 14 = 7 + 14$ Add 14 to both sides
$$4x = 21$$
$$x = \frac{21}{4} = 5\frac{1}{4}$$ Divide both sides by 4

Example

Solve $6x + 5 = 2(2x + 1)$

Grade D

$$6x + 5 = 2(2x + 1)$$
Expand the bracket first. ⟶ $6x + 5 = 4x + 2$

Then use the balancing method. ⟶ $6x - 4x + 5 = 4x - 4x + 2$ Take $4x$ from both sides
$$2x + 5 = 2$$
$$2x + 5 - 5 = 2 - 5$$ Take 5 from both sides
$$2x = -3$$
$$x = -\frac{3}{2} = -1\frac{1}{2}$$ Divide both sides by 2

Practice

Grade F   1   Solve $x + 9 = 0$

Grade E   2   Solve $2x - 7 = 6$

Grade D   3   Solve $7x - 1 = x - 7$

4   Solve $2(5x - 9) = 27$ **Grade D**

5   Solve $3(6h - 3) = 4(5h - 4)$ **Grade C**

6   Solve $\dfrac{3d}{2} + 4 = 27$ **Grade C**

Check your answers on page 169. For full worked solutions see the CD.
See the Student Book on the CD if you need more help.

Question	1	2	3	4	5	6
Grade	F	E	D	D	C	C
Student Book pages	U3 66–67	U3 67–68	U3 68–69	U3 69–70	U3 69–70	U3 69–70

Solving non-linear equations and inequalities

- A **quadratic** equation has an x^2 term (and no higher power of x).

- Quadratic equations can have 0, 1 or 2 solutions.

Key words

square ☐	quadratic ☐
square root ☐	

Example   Solve $4x^2 - 7 = 93$ Grade C

Use the balancing method to get the x^2 term on its own.

$$4x^2 - 7 = 93$$
$$4x^2 - 7 + 7 = 93 + 7 \quad \text{Add 7 to both sides}$$
$$4x^2 = 100$$
$$x^2 = 25 \quad \text{Divide both sides by 4}$$

Find the **square root.**

$$x = \pm 5, \ x = 5 \text{ or } x = -5$$

TIP
When you find a square root there are two solutions: one is positive, one is negative.

- You can find approximate solutions of more complicated equations by **trial and improvement**.

- A **cubic** equation has an x^3 term (and no higher power of x).

Key words

cube ☐	cubic ☐
trial and improvement	☐

Example   The equation $2x^3 + 3x = 400$ has a solution between 5 and 6. Use a trial and improvement method to find this solution. Give your answer correct to one decimal place.   Grade C

Work out $2x^3 + 3x$ using different values of x, until you get near to 400. Start with the x values given.

When $x = 5$,	$2x^3 + 3x = 265$	Smaller than 400
When $x = 6$,	$2x^3 + 3x = 450$	Bigger than 400
When $x = 5.5$,	$2x^3 + 3x = 349.25$	Smaller than 400
When $x = 5.7$,	$2x^3 + 3x = 387.486$	Smaller than 400
When $x = 5.8$,	$2x^3 + 3x = 407.624$	Bigger than 400

TIP
Work out carefully:
$(2 \times 5 \times 5 \times 5) + (3 \times 5) = 265$

EXAMINER'S TIP
Always write down an accurate answer, with decimals if possible.

So x lies between 5.7 and 5.8. Try the half-way value.

When $x = 5.75$, $2x^3 + 3x = 397.468$ Smaller than 400

The solution is between $x = 5.75$ and $x = 5.8$
Any number in this range rounds to 5.8 (to 1 d.p.).
So $x = 5.8$ (to 1 d.p.)

EXAMINER'S TIP
Remember to write down the solution to the equation. This is the value of x (5.8), not the number calculated (397.468).

- An **inequality** is a statement with one of these signs.

  $>$ means **greater than** $\geq$ means **greater than or equal to**

  $<$ means **less than** $\leq$ means **less than or equal to**

Key words

inequality ☐   integer ☐

- You can solve inequalities using the balancing method.

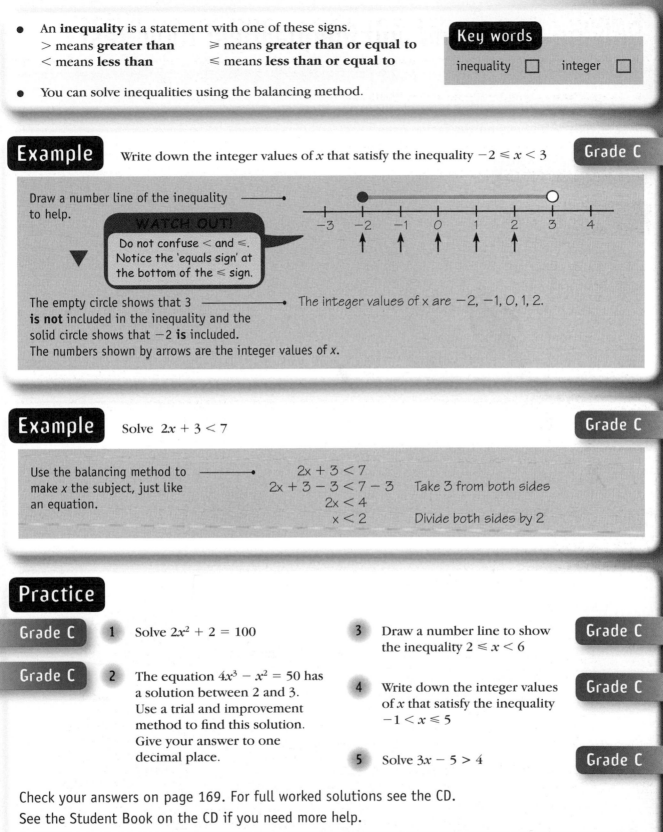

Example

Write down the integer values of x that satisfy the inequality $-2 \leq x < 3$ **Grade C**

Draw a number line of the inequality to help.

WATCH OUT!

Do not confuse $<$ and $\leq$. Notice the 'equals sign' at the bottom of the $\leq$ sign.

The empty circle shows that 3 **is not** included in the inequality and the solid circle shows that -2 **is** included.
The numbers shown by arrows are the integer values of x.

The integer values of x are $-2, -1, 0, 1, 2$.

Example

Solve $2x + 3 < 7$ **Grade C**

Use the balancing method to make x the subject, just like an equation.

$$2x + 3 < 7$$
$$2x + 3 - 3 < 7 - 3 \quad \text{Take 3 from both sides}$$
$$2x < 4$$
$$x < 2 \quad \text{Divide both sides by 2}$$

Practice

Grade C   **1**   Solve $2x^2 + 2 = 100$

Grade C   **2**   The equation $4x^3 - x^2 = 50$ has a solution between 2 and 3. Use a trial and improvement method to find this solution. Give your answer to one decimal place.

3   Draw a number line to show the inequality $2 \leq x < 6$ **Grade C**

4   Write down the integer values of x that satisfy the inequality $-1 < x \leq 5$ **Grade C**

5   Solve $3x - 5 > 4$ **Grade C**

Check your answers on page 169. For full worked solutions see the CD.
See the Student Book on the CD if you need more help.

Question	1	2	3	4	5
Grade	C	C	C	C	C
Student Book pages	U3 74–75	U3 75–76	U3 77–78	U3 78–81	U3 78–81

Solving equations and inequalities: topic test

Check how well you know this topic by answering these questions.
First cover the answers on the facing page.

Test questions

1 Solve

$$x + 3 = 7$$

2 Solve

$$4x - 3 = 21$$

3 Solve

$$5x + 2 = -4 + 3x$$

4 Solve

$$7x + 13 = 3(x + 5)$$

5 Solve

$$\frac{x}{3} + 2 = 1$$

6 Use a trial and improvement method to find a solution to

$$3x^3 + x^2 = 53$$

Give your answer correct to one decimal place.

7 **(a)** Write down the inequality that is shown on this number line:

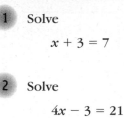

(b) Copy the number line below and show the inequality $-2 < x \leqslant 3$

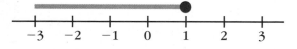

8 Write down the integer values of x that satisfy the inequality

$$-3 < x \leqslant 2$$

Now check your answers – see the facing page.

Cover this page while you answer the test questions opposite.

Worked answers

Revise this on...

F   **1**

$$x + 3 = 7$$
$$x + 3 - 3 = 7 - 3 \qquad \text{Subtract 3 from both sides}$$
$$x = 4$$

page 126

E   **2**

$$4x - 3 = 21$$
$$4x = 24 \qquad \text{Add 3 to both sides}$$
$$x = 6 \qquad \text{Divide both sides by 4}$$

page 126

D   **3**

$$5x + 2 = -4 + 3x$$
$$2x + 2 = -4 \qquad \text{Subtract 3x from both sides}$$
$$2x = -6 \qquad \text{Subtract 2 from both sides}$$
$$x = -3 \qquad \text{Divide both sides by 2}$$

page 126

C   **4**

$$7x + 13 = 3(x + 5)$$
$$7x + 13 = 3x + 15 \qquad \text{Expand the bracket}$$
$$4x + 13 = 15 \qquad \text{Subtract 3x from both sides}$$
$$4x = 2 \qquad \text{Subtract 13 from both sides}$$
$$x = \tfrac{2}{4} = \tfrac{1}{2} \qquad \text{Divide both sides by 4}$$

page 127

C   **5**

$$\tfrac{x}{3} + 2 = 1$$
$$\tfrac{x}{3} = -1 \qquad \text{Subtract 2 from both sides}$$
$$x = -3 \qquad \text{Multiply both sides by 3}$$

page 126

C   **6**

When $x = 1$,	$3x^3 + x^2 = 4$	Smaller than 53
When $x = 2$,	$3x^3 + x^2 = 28$	Smaller than 53
When $x = 3$,	$3x^3 + x^2 = 90$	Bigger than 53
When $x = 2.5$,	$3x^3 + x^2 = 53.125$	Bigger than 53
When $x = 2.4$,	$3x^3 + x^2 = 47.232$	Smaller than 53
When $x = 2.45$,	$3x^3 + x^2 = 50.120\,875$	Smaller than 53

The solution is between 2.45 and 2.5. Any number in this range rounds to 2.5 (to 1 d.p.).
So $x = 2.5$ (to 1 d.p.)

page 128

C   **7**

(a) $x \leqslant 1$

(b)

page 129

C   **8**

$-3 < x$ means the numbers are greater than -3 (not including -3).
$x \leqslant 2$ means the numbers are less than or equal to 2.
So the integer values are $-2, -1, 0, 1, 2$.

page 129

Tick the questions you got right.

Question	1	2	3	4	5	6	7	8
Grade	F	E	D	C	C	C	C	C

Mark the grade you are working at on your revision planner on page xii.

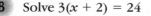

Algebra: subject test

Exam practice questions

1   Solve $x + 4 = 9$

2   The cost of buying a number of garden chairs over the internet is given by the formula

   total cost = number of chairs × £8.50 + delivery charge

   Work out the total cost for 6 chairs, with a delivery charge of £15.

3   At a shop, CDs cost £3 and DVDs cost £5.
   T is the total cost, in £, of buying c CDs and d DVDs.
   Write down a formula for T in terms of c and d.

4   Solve $2x - 3 = 15$

5   This is part of the travel graph for a motorist.

   (a)   How far from home is the motorist at 4 pm?

   (b)   For how long was the motorist stationary?

   (c)   What was the average speed, in km/h, of the motorist after 3 pm?

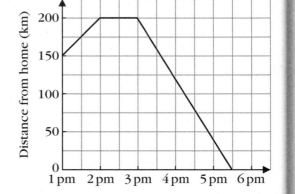

6   Simplify $q \times q \times q$

7   **(a)** Write down an expression, in terms of x, for the total of the angles in the triangle.
   (b) Write down an equation in terms of x.
   (c) Solve your equation to find the value of x.

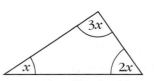

8   Solve $3(x + 2) = 24$

9   Expand $x(2x - 3y)$

10   Solve $4x - 1 = 2(x + 3)$

11 $-3 \leqslant n < 4$, where n is an integer.
Write down all the possible values of n.

12   The equation $3x^3 - x^2 = 58$ has a solution between 2 and 3.
Use a trial and improvement method to find this solution.
Give your answer correct to one decimal place.

13   **(a)** Complete the table of values for $y = 2x^2 - 3x + 2$.

x	-2	-1	0	1	2	3
y						

(b) On the grid, use your table of values to draw the graph of $y = 2x^2 - 3x + 2$.

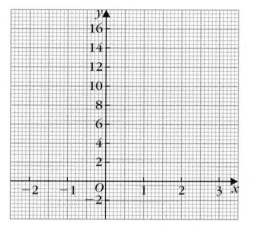

(c) Write down the minimum value of y.

(d) Use your graph to find
   (i) the value of y when $x = 2.5$
   (ii) the values of x when $y = 6$

14   Simplify
   (a) $\dfrac{8x^7}{2x^2}$ **(b)** $\dfrac{16y^5}{4y^2}$

Check your answers on page 169. For full worked solutions see the CD.

Tick the questions you got right.

Question	1	2	3	4	5ab	5c	6	7	8	9	10	11	12	13	14
Grade	F	E	E	E	E	D	D	D	D	C	C	C	C	C	B
Revise this on page	126	120	121	126	112		110	126	127	126	127	129	128	116	110–111

Mark the grade you are working at on your revision planner on page xii.

Go to the pages shown to revise for the ones you got wrong.

Polygons

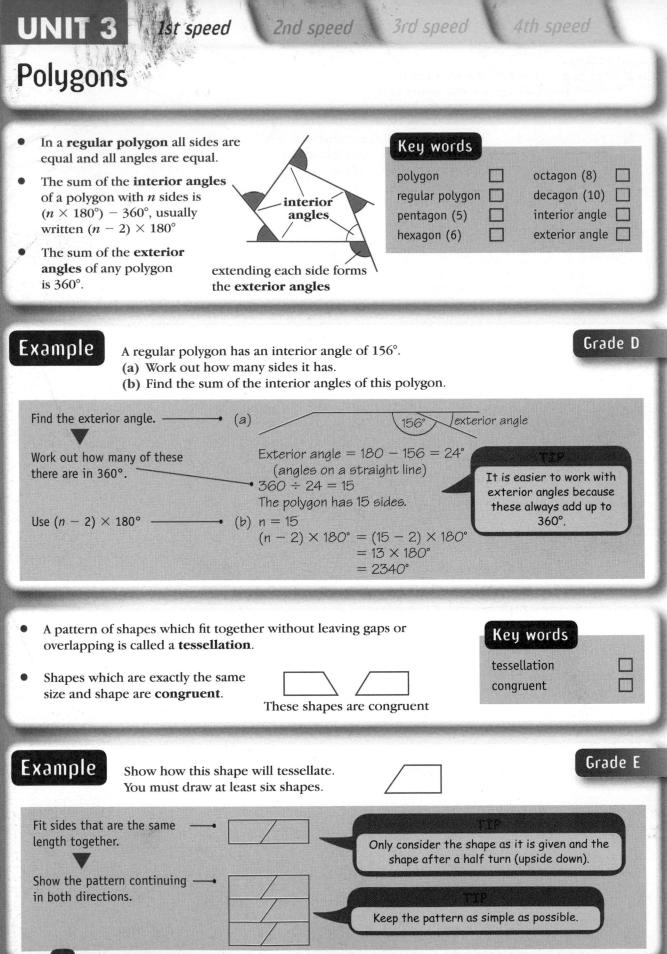

- In a **regular polygon** all sides are equal and all angles are equal.

- The sum of the **interior angles** of a polygon with n sides is $(n \times 180°) - 360°$, usually written $(n - 2) \times 180°$

- The sum of the **exterior angles** of any polygon is 360°.

interior angles

extending each side forms the **exterior angles**

Key words

polygon	☐	octagon (8)	☐
regular polygon	☐	decagon (10)	☐
pentagon (5)	☐	interior angle	☐
hexagon (6)	☐	exterior angle	☐

Example

Grade D

A regular polygon has an interior angle of 156°.

(a) Work out how many sides it has.

(b) Find the sum of the interior angles of this polygon.

Find the exterior angle.

▼

Work out how many of these there are in 360°.

Use $(n - 2) \times 180°$

(a)   156°   exterior angle

Exterior angle = 180 − 156 = 24°
  (angles on a straight line)
360 ÷ 24 = 15
The polygon has 15 sides.

(b) $n = 15$
$(n - 2) \times 180° = (15 - 2) \times 180°$
$= 13 \times 180°$
$= 2340°$

TIP

It is easier to work with exterior angles because these always add up to 360°.

- A pattern of shapes which fit together without leaving gaps or overlapping is called a **tessellation**.

- Shapes which are exactly the same size and shape are **congruent**.

These shapes are congruent

Key words

tessellation	☐
congruent	☐

Example

Grade E

Show how this shape will tessellate.
You must draw at least six shapes.

Fit sides that are the same length together.

▼

Show the pattern continuing in both directions.

TIP

Only consider the shape as it is given and the shape after a half turn (upside down).

TIP

Keep the pattern as simple as possible.

Practice

1 Write down the letters of the shapes that are congruent.

Grade F

2 Show how these shapes will tessellate.

(a) 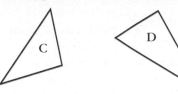 **(b)**

Grade F

3 Find the size of the angles marked *a* and *b*.

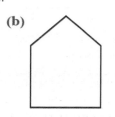

Grade E

4 *QRST* is a square.

PQT is an equilateral triangle.

(a) Show that angle *PTS* is 150°.

(b) Work out the size of angle *PST*.

(c) Show that angle *PXQ* is 75°.

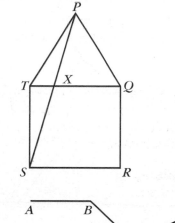

Grade D

5 *ABCD* is part of a regular octagon.
DCE is part of a regular hexagon.

(a) Work out the size of angle *BCE*.

(b) Explain why angle *CBE* = angle *CEB*.

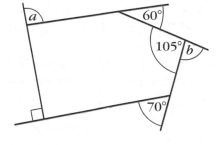

Grade C

Check your answers on pages 169–170. For full worked solutions see the CD.
See the Student Book on the CD if you need more help.

Question	1	2	3	4	5
Grade	F	F	E	D	C
Student Book pages	U3 175–177	U3 183–184	U3 180–182	U3 178–179	U3 180–182

Drawing and calculating

- A **bearing** is the angle measured from facing North and turning clockwise. It is always a three-figure number.

N

The angle is measured clockwise from the North.

Kim 123°

The bearing of the ship from Kim is 123°

- The **locus** of points equidistant from **one point** is a circle.

×

- The locus of points equidistant from **two points** is the **perpendicular bisector** of the line joining the two points.

×   ×

- The locus of points equidistant from **two lines** is made by bisecting the angle formed where they meet.

X
A   Y

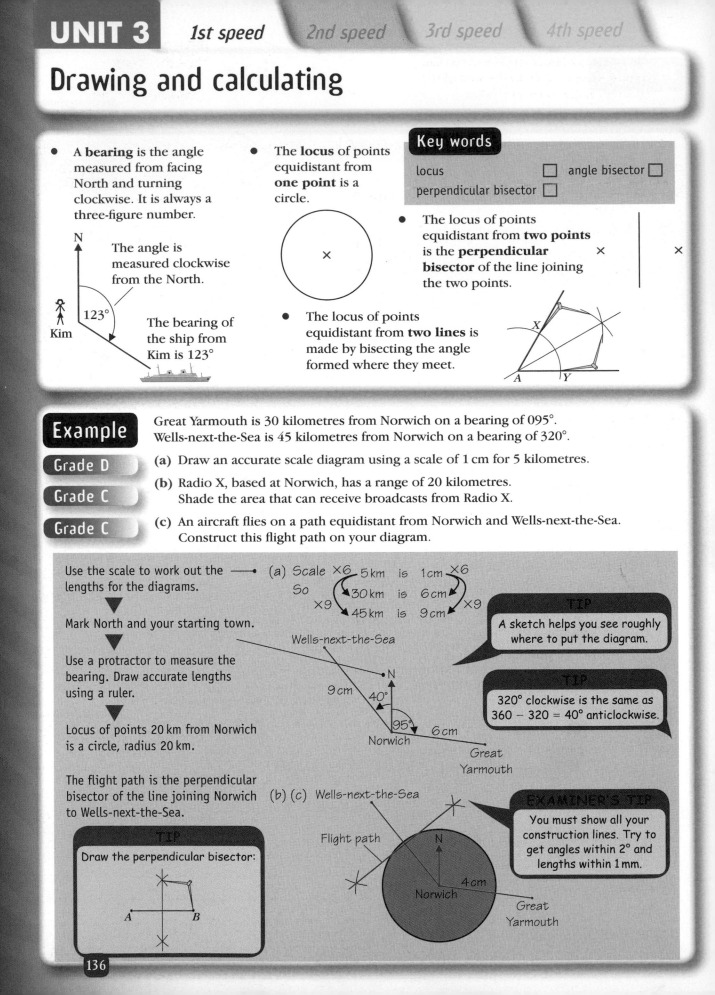

Key words

locus ☐   angle bisector ☐
perpendicular bisector ☐

Example

Grade D

Grade C

Grade C

Great Yarmouth is 30 kilometres from Norwich on a bearing of 095°.
Wells-next-the-Sea is 45 kilometres from Norwich on a bearing of 320°.

(a) Draw an accurate scale diagram using a scale of 1 cm for 5 kilometres.

(b) Radio X, based at Norwich, has a range of 20 kilometres.
Shade the area that can receive broadcasts from Radio X.

(c) An aircraft flies on a path equidistant from Norwich and Wells-next-the-Sea.
Construct this flight path on your diagram.

Use the scale to work out the lengths for the diagrams.

▼

Mark North and your starting town.

▼

Use a protractor to measure the bearing. Draw accurate lengths using a ruler.

▼

Locus of points 20 km from Norwich is a circle, radius 20 km.

The flight path is the perpendicular bisector of the line joining Norwich to Wells-next-the-Sea.

TIP

Draw the perpendicular bisector:

A   B

(a) Scale ×6 5 km is 1 cm ×6
So   ×9 30 km is 6 cm ×9
      ×9 45 km is 9 cm ×9

Wells-next-the-Sea

9 cm   N
        40°
        95°   6 cm
Norwich
        Great
        Yarmouth

(b) (c) Wells-next-the-Sea

Flight path   N

        4 cm
Norwich
        Great
        Yarmouth

TIP

A sketch helps you see roughly where to put the diagram.

TIP

320° clockwise is the same as 360 − 320 = 40° anticlockwise.

EXAMINER'S TIP

You must show all your construction lines. Try to get angles within 2° and lengths within 1 mm.

- **Pythagoras' theorem** states that in a right-angled triangle the square on the **hypotenuse** is equal to the sum of the squares on the other two sides.

$$c^2 = a^2 + b^2 \quad \text{or} \quad a^2 + b^2 = c^2$$

Key words

Pythagoras' theorem ☐
hypotenuse ☐

Grade C

Example   Find the length of *AB*.

TIP
The hypotenuse is the longest side = 18 cm

Use Pythagoras' theorem. ⟶ $18^2 = 12^2 + AB^2$

WATCH OUT!
Don't forget to take the square root.

Rearrange to make the ⟶ $AB^2 = 18^2 - 12^2 = 180$
unknown side the subject.

$AB = \sqrt{180} = 13.4 \text{ cm}$

TIP
Always check that your answer leaves the hypotenuse as the longest side.

TIP
For more on rearranging formulae see page 122.

Practice

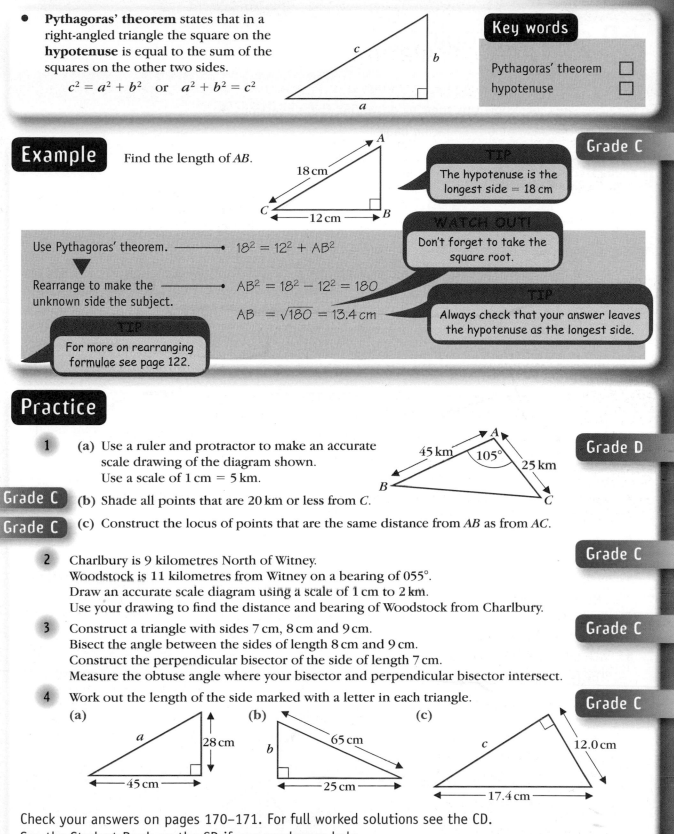

1   (a) Use a ruler and protractor to make an accurate scale drawing of the diagram shown. Use a scale of 1 cm = 5 km.

Grade D

(b) Shade all points that are 20 km or less from *C*.   **Grade C**

(c) Construct the locus of points that are the same distance from *AB* as from *AC*.   **Grade C**

2   Charlbury is 9 kilometres North of Witney.
Woodstock is 11 kilometres from Witney on a bearing of 055°.
Draw an accurate scale diagram using a scale of 1 cm to 2 km.
Use your drawing to find the distance and bearing of Woodstock from Charlbury.

Grade C

3   Construct a triangle with sides 7 cm, 8 cm and 9 cm.
Bisect the angle between the sides of length 8 cm and 9 cm.
Construct the perpendicular bisector of the side of length 7 cm.
Measure the obtuse angle where your bisector and perpendicular bisector intersect.

Grade C

4   Work out the length of the side marked with a letter in each triangle.

Grade C

(a) a, 28 cm, 45 cm

(b)   65 cm, b, 25 cm

(c) c, 12.0 cm, 17.4 cm

Check your answers on pages 170–171. For full worked solutions see the CD.
See the Student Book on the CD if you need more help.

Question	1a	1bc	2	3	4
Grade	D	C	C	C	C
Student Book pages	U3 154	U3 166–168	U3 157–160	U3 160–166	U3 205–211

2-D shapes: topic test

Check how well you know this topic by answering these questions.
First cover the answers on the facing page.

Test questions

1 Write the mathematical name for each of these shapes.

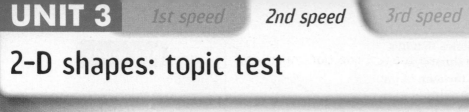

(a) **(b)** **(c)** **(d)**

2 Work out the sum of the interior angles of an irregular pentagon.

3 Write down the letters of the shape(s) that are congruent to shape A.

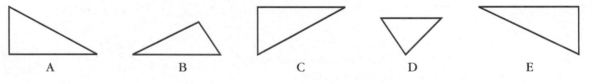

A   B   C   D   E

4 The exterior angle of a regular polygon is 15°. Work out how many sides this polygon has.

5 The bearing of *B* from *A* is 056°. Write down the bearing of *A* from *B*.

6 Mark two points, *A* and *B*, and construct the locus of the points that are equidistant from *A* and *B*.

7 Draw an angle of about 50°. Construct the bisector of this angle.

8 *A*, *B* and *C* are three towns.
The bearing of *B* from *A* is 070° and the distance is 50 km.
The bearing of *C* from *B* is 130° and the distance is 60 km.
Use a scale of 1 cm = 10 km to draw a scale diagram.
Use your diagram to find the distance and bearing of *A* from *C*.

9 Work out the length of **(a)** *AD* **(b)** *AC*.

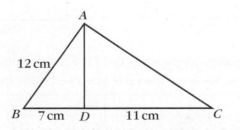

Now check your answers – see the facing page.

Cover this page while you answer the test questions opposite.

Worked answers

Revise this on...

G 1 (a) Kite (b) Hexagon (c) Parallelogram (d) Trapezium page 134

F 2 Angle sum = $(n - 2) \times 180°$ page 134
$n = 5$ so angle sum = $3 \times 180° = 540°$

F 3 Shape C page 134

D 4 Sum of the exterior angles is 360°. page 134
Number of sides = $360 \div 15 = 24$

D 5 $056 + 180 = 236°$ (The bearing is in the opposite direction so page 136
a half turn of ±180° is used.)

D 6 page 136

D 7

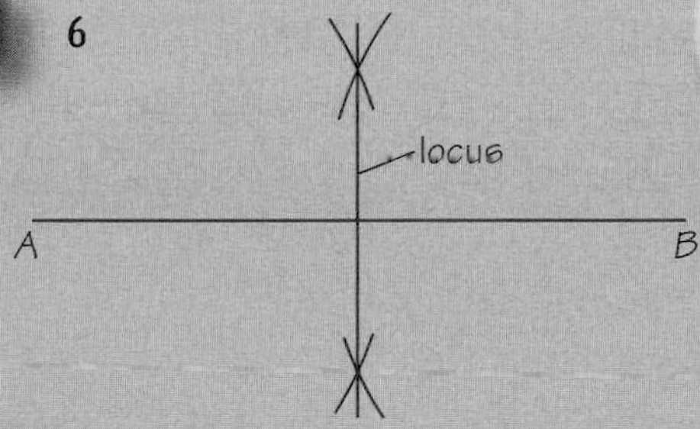

locus

A   B

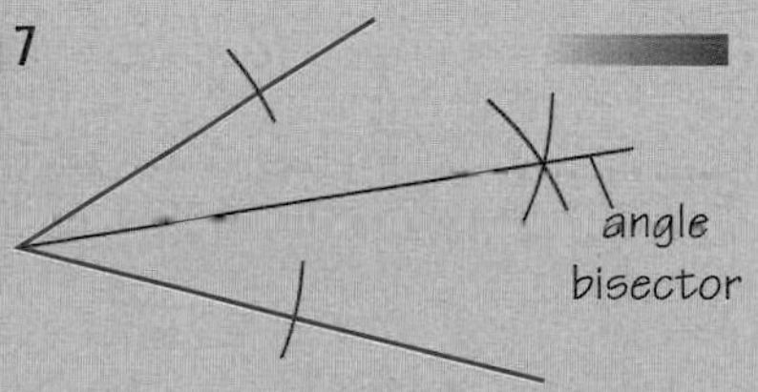

angle bisector

D 8 page 136

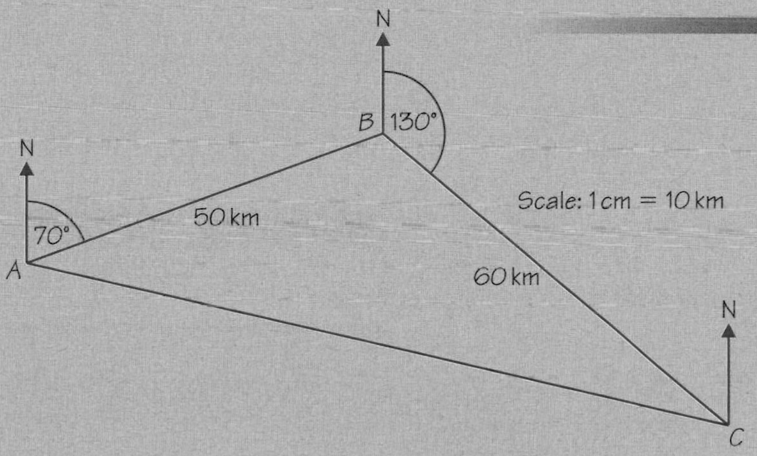

N

B 130°

N

50 km   Scale: 1 cm = 10 km

70°
A   60 km

N

C

The bearing of A from C is 283° and the distance is 95 km.

C 9 (a) $AD^2 + 7^2 = 12^2$ so $AD^2 = 12^2 - 7^2 = 95$, $AD = \sqrt{95} = 9.75$ cm page 137

(b) $AC^2 = AD^2 + CD^2 = 95 + 11^2 = 216$ so $AC = \sqrt{216} = 14.7$ cm

Tick the questions you got right.

Question	1	2	3	4	5	6	7	8	9
Grade	G	F	F	D	D	D	D	D	C

Mark the grade you are working at on your revision planner on page xii.

3-D shapes

- A **net** is a 2-D shape that can be folded into a 3-D shape.

- A **prism** is a shape which has a uniform cross-section.

- A 3-D shape has a **plane of symmetry** if the plane divides the shape into two halves and one half is the mirror image of the other.

Key words

net ☐   plane of symmetry ☐
prism ☐

Example   Draw the net for this 3-D shape.   **Grade D**

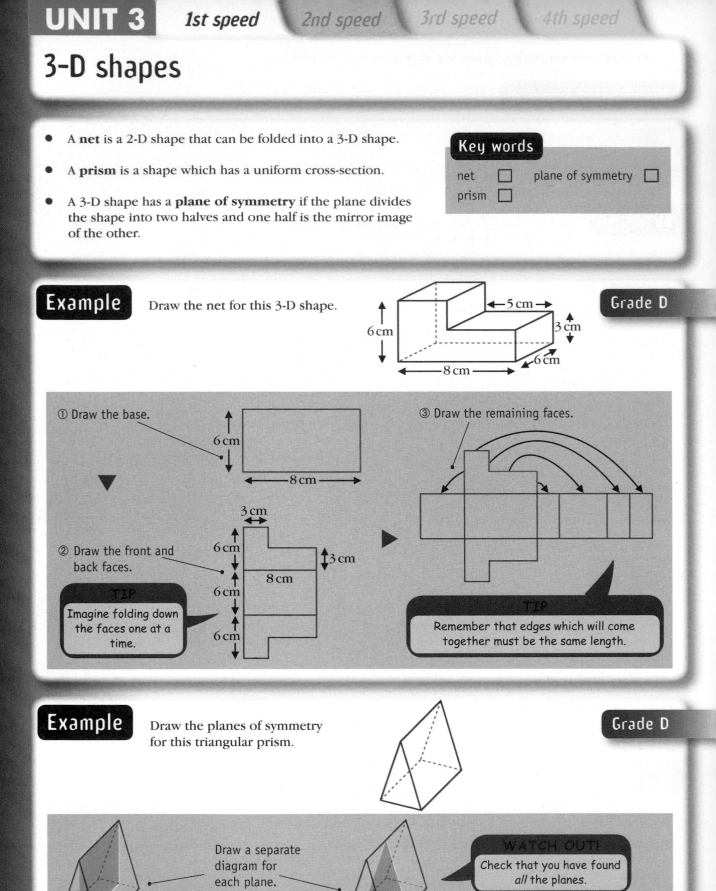

① Draw the base.

6 cm
8 cm

TIP
Imagine folding down the faces one at a time.

② Draw the front and back faces.

3 cm
6 cm   3 cm
8 cm
6 cm
6 cm

③ Draw the remaining faces.

TIP
Remember that edges which will come together must be the same length.

Example   Draw the planes of symmetry for this triangular prism.   **Grade D**

Draw a separate diagram for each plane.

WATCH OUT!
Check that you have found *all* the planes.

- The **plan** of a solid is the view when seen from above.

- The **front elevation** is the view when seen from the front.

- The **side elevation** is the view when seen from the side.

Example This diagram on isometric paper shows a 3-D shape made from nine cubes. Draw the plan and elevations for this shape

Grade D

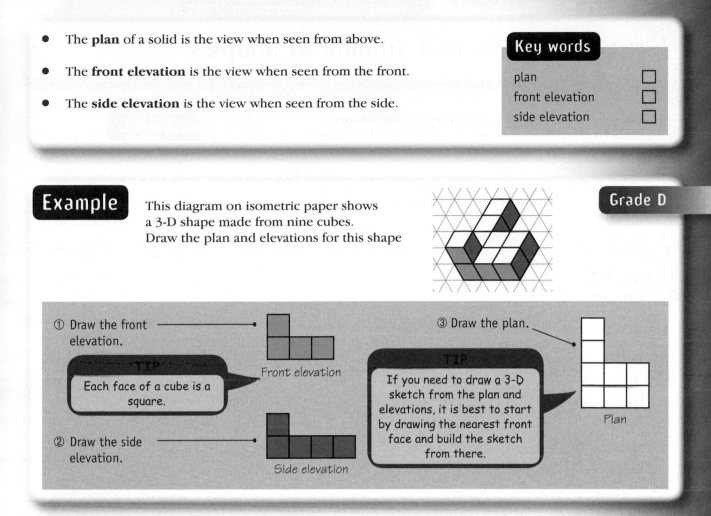

① Draw the front elevation.

TIP
Each face of a cube is a square.

Front elevation

② Draw the side elevation.

Side elevation

③ Draw the plan.

TIP
If you need to draw a 3-D sketch from the plan and elevations, it is best to start by drawing the nearest front face and build the sketch from there.

Plan

Practice

1 Here is a diagram of a 3-D shape.

(a) Draw the plane of symmetry for the shape.

(b) Construct the net for the shape.

(c) Sketch the plan and elevations for the shape.

Grade D

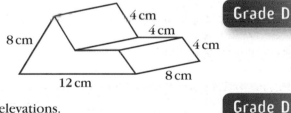

8 cm
4 cm
4 cm
4 cm
12 cm
8 cm

2 Draw a sketch of the 3-D prism with this plan and elevations.

Grade D

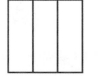 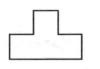 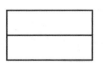

Check your answers on page 171. For full worked solutions see the CD.
See the Student Book on the CD if you need more help.

Question	1a	1b	1c	2
Grade	D	D	D	D
Student Book pages	U3 144–145	U3 168–172	U3 173–175	U3 173–175

Perimeter, area and volume of shapes

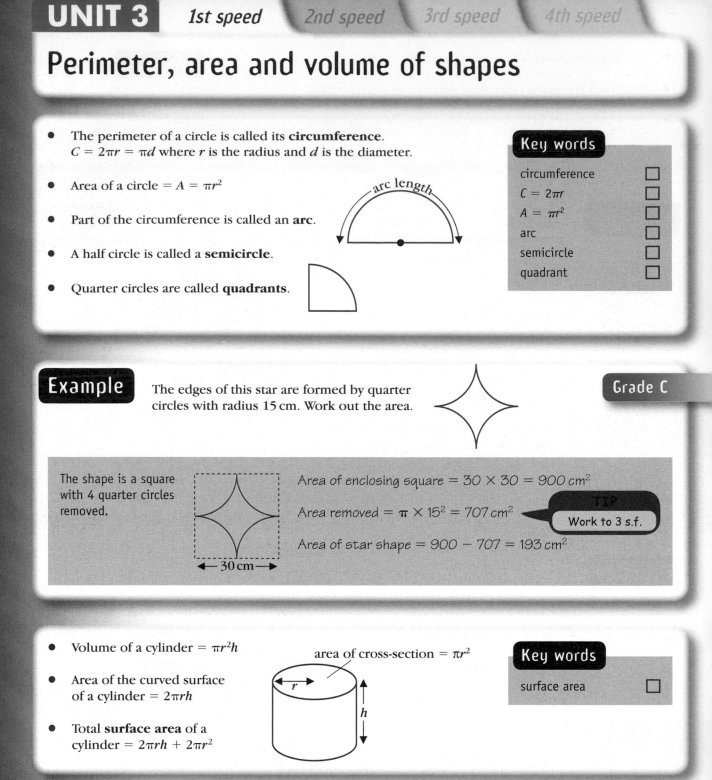

- The perimeter of a circle is called its **circumference**.
  $C = 2\pi r = \pi d$ where r is the radius and d is the diameter.

- Area of a circle = $A = \pi r^2$

- Part of the circumference is called an **arc**.

arc length

- A half circle is called a **semicircle**.

- Quarter circles are called **quadrants**.

Key words

circumference	☐
$C = 2\pi r$	☐
$A = \pi r^2$	☐
arc	☐
semicircle	☐
quadrant	☐

Example   The edges of this star are formed by quarter circles with radius 15 cm. Work out the area.   **Grade C**

The shape is a square with 4 quarter circles removed.

Area of enclosing square = $30 \times 30 = 900 \, \text{cm}^2$

Area removed = $\pi \times 15^2 = 707 \, \text{cm}^2$

TIP Work to 3 s.f.

Area of star shape = $900 - 707 = 193 \, \text{cm}^2$

$\leftarrow 30\,\text{cm} \rightarrow$

- Volume of a cylinder = $\pi r^2 h$

- Area of the curved surface of a cylinder = $2\pi rh$

- Total **surface area** of a cylinder = $2\pi rh + 2\pi r^2$

area of cross-section = πr^2

r

h

Key words

surface area	☐

Example   A washer is made from a circular disc of metal, diameter 30 mm, with a circular hole of radius 5 mm.   **Grade C**

The washer is 2 mm thick.

The density of the metal is $7.8 \, \text{g/cm}^3$.

Work out the mass of 50 of these washers.

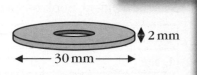

2 mm

$\leftarrow 30\,\text{mm} \rightarrow$

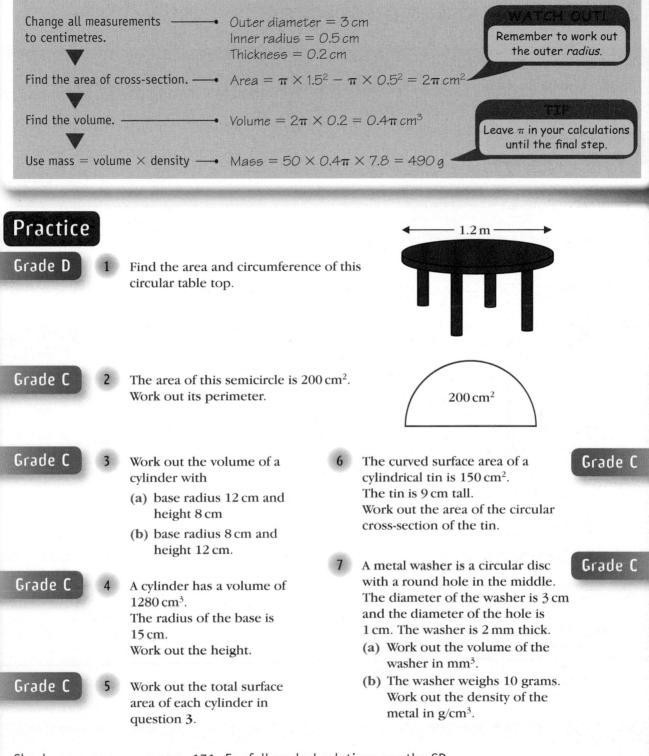

Change all measurements to centimetres.

Outer diameter = 3 cm
Inner radius = 0.5 cm
Thickness = 0.2 cm

WATCH OUT!
Remember to work out the outer *radius*.

Find the area of cross-section.

Area = $\pi \times 1.5^2 - \pi \times 0.5^2 = 2\pi$ cm^2

Find the volume.

Volume = $2\pi \times 0.2 = 0.4\pi$ cm^3

TIP
Leave π in your calculations until the final step.

Use mass = volume × density

Mass = $50 \times 0.4\pi \times 7.8 = 490$ g

Practice

Grade D **1** Find the area and circumference of this circular table top.

1.2 m

Grade C **2** The area of this semicircle is 200 cm^2. Work out its perimeter.

200 cm^2

Grade C **3** Work out the volume of a cylinder with

(a) base radius 12 cm and height 8 cm

(b) base radius 8 cm and height 12 cm.

Grade C **4** A cylinder has a volume of 1280 cm^3. The radius of the base is 15 cm. Work out the height.

Grade C **5** Work out the total surface area of each cylinder in question **3**.

6 The curved surface area of a cylindrical tin is 150 cm^2. The tin is 9 cm tall. Work out the area of the circular cross-section of the tin. **Grade C**

7 A metal washer is a circular disc with a round hole in the middle. The diameter of the washer is 3 cm and the diameter of the hole is 1 cm. The washer is 2 mm thick. **Grade C**

(a) Work out the volume of the washer in mm^3.

(b) The washer weighs 10 grams. Work out the density of the metal in g/cm^3.

Check your answers on page 171. For full worked solutions see the CD.
See the Student Book on the CD if you need more help.

Question	1	2	3	4	5	6	7a	7b
Grade	D	C	C	C	C	C	C	C
Student Book pages	U3 194–198	U3 194–198	U3 203–205 U2 169–170	U3 203–205 U2 169–170	U3 203–205 U2 169–170	U3 203–205 U2 169–170	U3 203–205	U2 148–149

2-D and 3-D shapes: topic test

Check how well you know this topic by answering these questions.
First cover the answers on the facing page.

Test questions

1 Draw the two planes of symmetry in this solid shape.

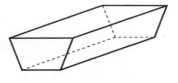

2 Draw a sketch of the solid shape made from this net.

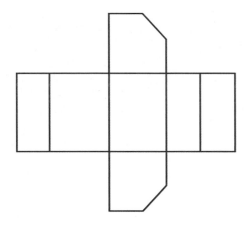

3 Sketch the plan and elevations for this prism.

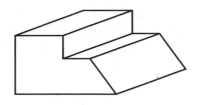

4 The area of a circle is 30.6 cm².
Work out

   (a) the radius

   (b) the circumference.

5 Work out the volume of a cylinder with base diameter 16 cm and height 40 cm.

Now check your answers – see the facing page.

Cover this page while you answer the test questions opposite.

Worked answers

Revise this on...

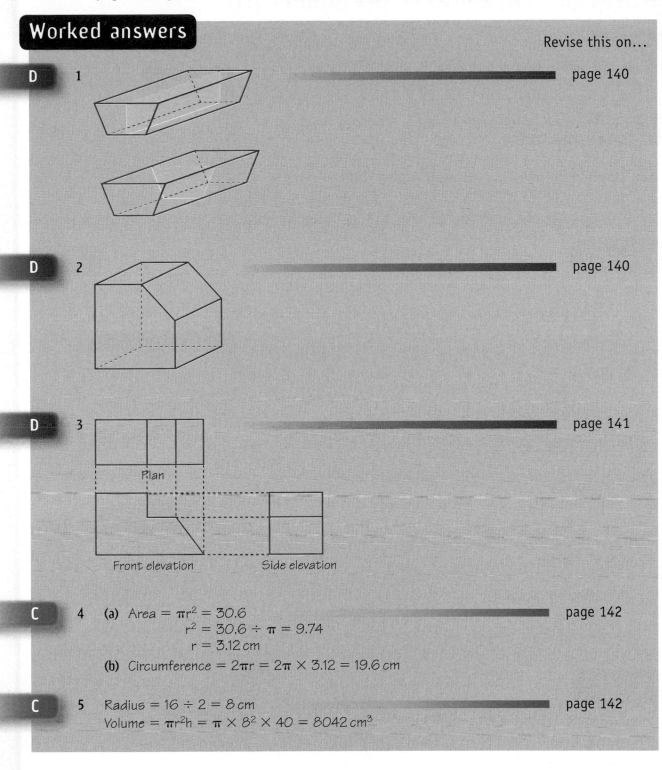

D **1**
page 140

D **2**
page 140

D **3**
Plan

Front elevation   Side elevation
page 141

C **4** **(a)** Area = πr^2 = 30.6

r^2 = 30.6 ÷ π = 9.74

r = 3.12 cm

(b) Circumference = $2\pi r$ = 2π × 3.12 = 19.6 cm
page 142

C **5** Radius = 16 ÷ 2 = 8 cm

Volume = $\pi r^2 h$ = π × 8^2 × 40 = 8042 cm³
page 142

Tick the questions you got right.

Question	1	2	3	4	5
Grade	D	D	D	C	C

Mark the grade you are working at on your revision planner on page xii.

Rotation, reflection and symmetry

- A 2-D shape has a **line of symmetry** if the line divides the shape into two halves and one half is the **mirror image** of the other half.

- A 2-D shape has **rotational symmetry** if the appearance of its starting position occurs two or more times during a full turn.

- The **order of rotational symmetry** is the number of times the original appearance occurs during a full turn.

Key words

line of symmetry ☐
mirror image ☐
rotational symmetry ☐
order of rotational symmetry ☐

Example

(a) Draw all the lines of symmetry on this shape.
(b) What is its order of rotational symmetry?

Grade E

Imagine folding the shape in half, ────• (a)
as many times as possible.

TIP
A regular polygon has the same number of lines of symmetry as sides.

Imagine rotating the shape. ────• (b) Rotational symmetry of order 5
How many times does it look the same by
the time it comes back to the begining?

- A **reflection** is the image formed by a mirror line called the line of reflection.

- To describe a reflection you need to give the equation of the line of symmetry.

- A **rotation** turns a shape through an angle about a fixed point (the **centre of rotation**).

- To describe a rotation you need to give the centre, the angle, and whether it is clockwise or anticlockwise.

Key words

reflection ☐
rotation ☐
centre of rotation ☐

Example

Describe fully the single transformation
which maps shape **P** on to **Q**.

Grade C

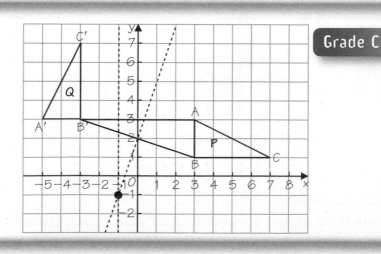

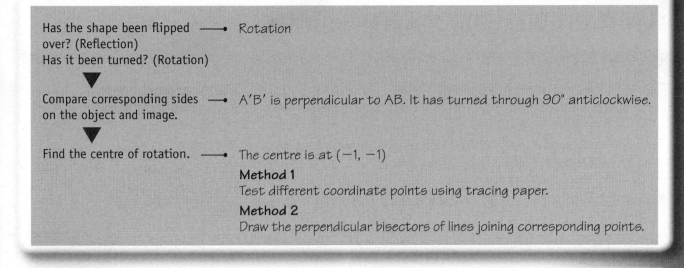

Has the shape been flipped over? (Reflection) ⟶ Rotation
Has it been turned? (Rotation)

▼

Compare corresponding sides on the object and image. ⟶ A'B' is perpendicular to AB. It has turned through 90° anticlockwise.

▼

Find the centre of rotation. ⟶ The centre is at (−1, −1)

Method 1
Test different coordinate points using tracing paper.

Method 2
Draw the perpendicular bisectors of lines joining corresponding points.

Practice

Grade D **1** Rotate the triangle by a half turn about the origin.

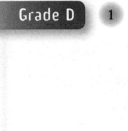

Grade D **2** (a) Reflect triangle **A** in the line x = 5. Label the image **C**.

Grade D (b) Describe fully the transformation that maps **A** on to **B**.

Grade C (c) Describe fully the single transformation that maps **C** on to **B**.

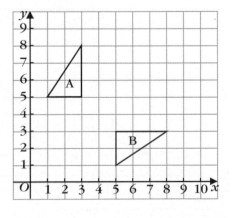

3 (a) Reflect **P** in the y-axis. Grade C
Label the image **Q**.

(b) Reflect **Q** in the line y = 2. Label the image **R**.

(c) Describe fully the single transformation that maps **P** on to **R**.

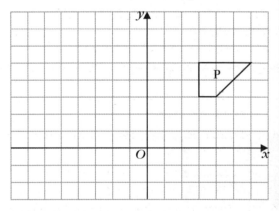

Check your answers on page 171.
For full worked solutions see the CD.
See the Student Book on the CD if you need more help.

Question	1		2ab	2c	3
Grade	D		D	C	C
Student Book pages		U3 132–134	U3 129–131 U3 142–144	U3 129–131 U3 142–144	

Translations and enlargements

- To describe a **translation** (sliding movement) you can give the horizontal (sideways) and vertical (up or down) movements or use a **column vector**.

- An **enlargement** changes the size but not the shape of an object. The **scale factor** of the enlargement is the value that the lengths of the original object are multiplied by.

- In an enlargement, image lines are parallel to their corresponding object lines.

- To describe an enlargement fully you need to give the scale factor and the **centre of enlargement**.

Key words

translation ☐
column vector ☐
enlargement ☐
scale factor ☐
centre of enlargement ☐
similar ☐

Example   Translate shape **A** by the vector $\begin{pmatrix} 2 \\ -3 \end{pmatrix}$. Label the image **B**.

Grade C

The 'top' number gives the horizontal movement. The 'bottom' number gives the vertical movement.

TIP
Vertical movement by −3 means move 3 squares **downwards**.

Example   Enlarge triangle *DEF* by a scale factor of 3 and with centre of enlargement (−3, 1).

Grade C

Mark the centre of enlargement *C*.

▼

Draw a line from *C* through each vertex. The image points will lie along these lines.

▼

Use the scale factor to identify the image points and join them.
Here, *C'D'* = 3 × *CD*

TIP
Always check that corresponding sides are parallel.

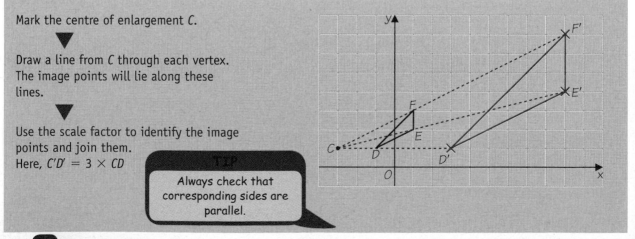

- A scale factor less than 1 means that the image is *smaller* than the object.

Example

Work out the scale factor of the enlargement that maps rectangle *ABCD* on to *FGHI*.

Grade E

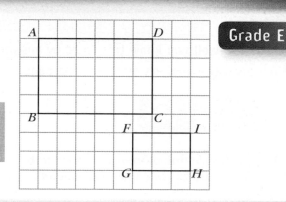

Compare corresponding line segments. → Scale factor $= \dfrac{FG}{AB} = \dfrac{2}{4} = \dfrac{1}{2}$

Practice

Grade E

1 Enlarge the shape using a scale factor of 3.

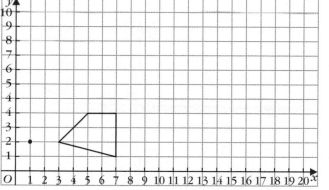

Grade C

2 Enlarge the shape using centre (1, 2) and a scale factor of 2.

3 **(a)** Translate the shape **R** by vector $\begin{pmatrix} 4 \\ 2 \end{pmatrix}$.

Label the image **S**.

Grade C

(b) Translate the shape **S** by vector $\begin{pmatrix} -3 \\ -5 \end{pmatrix}$.

Label the image **T**.

(c) Describe fully the single transformation that maps **T** on to **R**.

Check your answers on pages 171–172. For full worked solutions see the CD.
See the Student Book on the CD if you need more help.

Question	1	2	3
Grade	E	C	C
Student Book pages	U3 138–142	U3 138–142	U3 134–138, 142–144

Transformations: topic test

Check how well you know this topic by answering these questions.
First cover the answers on the facing page.

Test questions

1 **(a)** Shade **one** more square so that the shape has one line of symmetry.

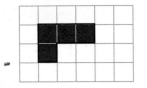

(b) Shade **one** more square so that the shape has rotational symmetry. What is the order of rotational symmetry?

2 **(a)** Translate shape **A** 5 units to the left and 2 units upwards. Label the image **X**.

(b) Translate shape **A** by the column vector $\begin{pmatrix} 2 \\ -3 \end{pmatrix}$. Label the image **Y**.

(c) Rotate shape **A** by a rotation of 90° clockwise with centre of rotation (1, 3). Label the image **Z**.

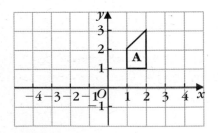

3 Enlarge shape **B** by a scale factor of 1.5 using (−3, 3) as the centre.

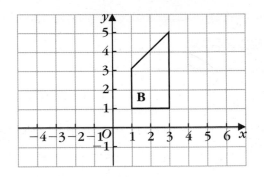

4 Reflect shape **C** in the line $x = y$.

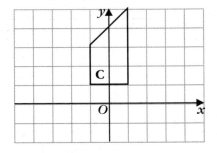

5 Describe fully the transformation which maps **D** on to **E**.

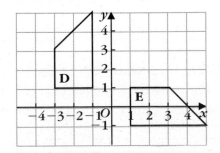

Now check your answers – see the facing page.

Cover this page while you answer the test questions opposite.

Worked answers

Revise this on...

E 1 (a) There are several correct answers:

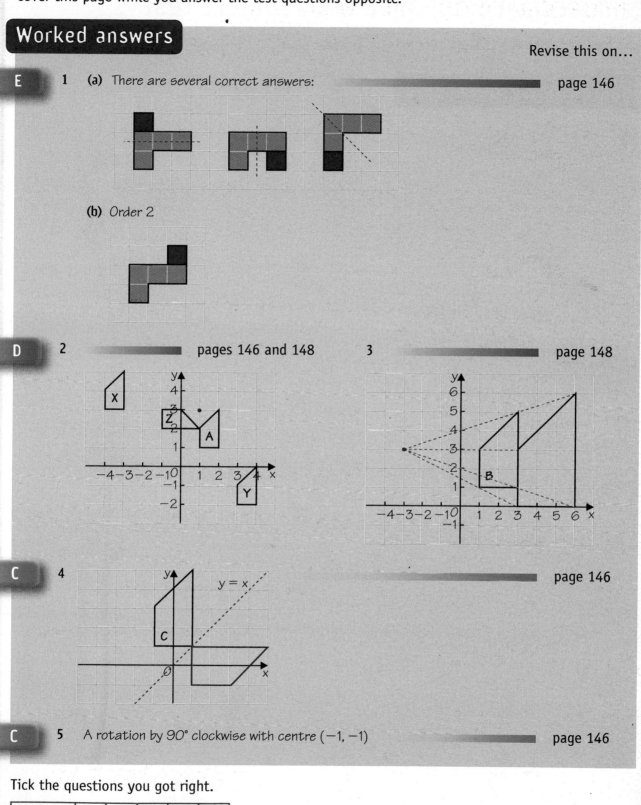

page 146

(b) Order 2

2 pages 146 and 148

3 page 148

C 4 page 146

C 5 A rotation by 90° clockwise with centre (−1, −1) page 146

Tick the questions you got right.

Question	1	2	3	4	5
Grade	E	D	D	C	C

Mark the grade you are working at on your revision planner on page xii.

Shape, space and measure: subject test

Exam practice questions

1 Pairs of shapes in the diagram are congruent. List these pairs.

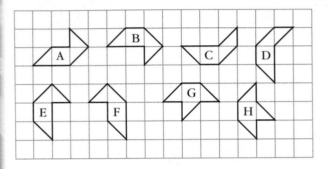

2 (a)

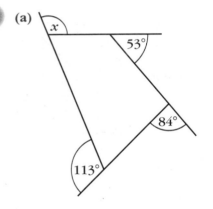

Find x

(b)

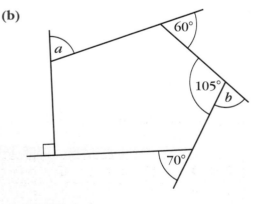

Find a

3 Copy and draw six shapes to show how the shape tessellates.

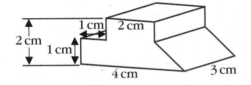

4 Sketch the plan and elevations for this shape.

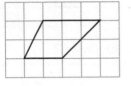

5 Sketch a net for this shape.

6 A wheel has radius 25 cm.
Work out how far the wheel travels with 500 revolutions.

7 (a) Translate shape A by $\begin{pmatrix} 3 \\ -2 \end{pmatrix}$

(b) Describe fully the transformation that maps shape A onto shape B.

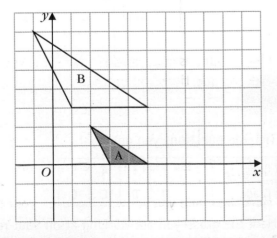

8 The dance floor shown in the diagram measures 10 m by 18 m.
A and *B* are two pillars.
Construct a diagram to show all the points on the dance floor that are more than 2.5 m from walls and pillars.
Use a scale of 1 cm to represent 1 m.

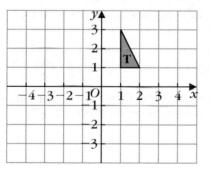

9 *A* is 8 km due North of *B*. A ship leaves *A* and travels on a bearing of 120°.
Another ship leaves *B* and travels on a bearing of 068°.
Using a scale of 1 cm to represent 1 km draw a scale drawing and use it to find how far from *A* the ships' paths cross.

10 (a) Reflect the triangle **T** in the line $y = x$. Label the shape **R**.

(b) Rotate the triangle **T** about $(0, 0)$ through 180°. Label the shape **S**.

(c) Describe fully the transformation that will move shape **S** on to shape **R**.

11 *ABCD* is part of a regular pentagon. Work out the size of angle BXC.

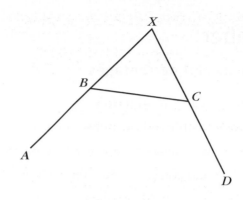

12 The area of a circle is 200 cm². Find the diameter of the circle.

13 *ABC* is a right-angled triangle with *AB* = 15.3 cm and *BC* = 25.5 cm. Work out the length of *AC*.

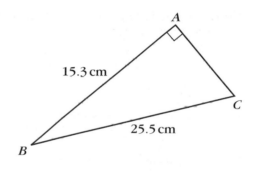

Check your answers on pages 172–173. For full worked solutions see the CD.
Tick the questions you got right.

Question	1	2	3	4	5	6	7	8	9	10	11	12	13
Grade	G	E	E	D	D	D	D	C	C	C	C	C	C
Revise this on page	134	134	134	141	140	142, 143	148, 149	136	136	146, 148	134	142, 143	134

Mark the grade you are working at on your revision planner on page xii.
Go to the pages shown to revise for the ones you got wrong.

Unit 3 Key points

Number

Fractions and percentages

- **Addition and subtraction**
  - You can only add and subtract fractions that have the same bottom number (denominator).
  - Start by dealing with any whole numbers.
  - Then find equivalent fractions with the same denominator for the fractions.
  - You can then add or subtract.

- **Multiplication and division**

  When you multiply fractions you write any mixed numbers as improper fractions. You then multiply the numerators and the denominators.

  When you divide fractions you write any mixed numbers as improper fractions. Then write down the first fraction, invert the second fraction and multiply.

- To compare fractions, decimals and percentages you can change them all to percentages.

- To find a percentage of an amount you can: change the percentage to a fraction and multiply *or* change the percentage to a decimal and multiply *or* work from 10%.

- To **increase** a number by a percentage, you find the percentage of that number and then add this to the starting number. To **decrease** a number by a percentage, you find the percentage of that number and then subtract this from the starting number.

- To write one number as a percentage of another write the amounts as a fraction, convert the fraction to a decimal and then change the decimal to a percentage by multiplying by 100.

- A **price index** shows how the price of something changes over time. The index always starts at 100. An index greater than 100 shows a price rise, while an index less than 100 shows a price fall.

Ratio and proportion

- Simplifying ratios
  - To **simplify** a ratio you divide both its numbers by a common factor.
  - When a ratio cannot be simplified it is said to be in its **lowest terms**.
  - Two ratios are **equivalent** when they both simplify to the same ratio.

- To share an amount in a given ratio find the total of all the ratio numbers, then split the amount into fractions with a denominator that is the total.

- Two quantities are in **direct proportion** if their ratio stays the same when the quantities increase or decrease.

Algebra

Algebra and graphs

- A graph representing a **linear relationship** is always a straight line.

- **Distance–time graphs** are used to relate the distance travelled to the time taken, and to calculate speeds.

- An equation containing an x^2-term is called a **quadratic equation**.

- The graph of a quadratic equation is a curved quadratic graph, or parabola. It has a symmetrical U-shape: $\cup$ or $\cap$.

Formulae

- A **word formula** uses words to represent a relationship between quantities. For example:

  pay = rate of pay $\times$ hours worked

- An **algebraic formula** uses letters to represent a relationship between quantities. For example, the perimeter of a rectangle, P, is related to its length l and width w by

  $P = 2l + 2w$

- The **subject** of a formula appears on its own on one side of the formula and does not appear on the other side. For example:

  $t = 4l + 4$ can be rearranged to give $l = \dfrac{t - 4}{4}$

  t is the subject l is the subject

Solving equations and inequalities

- In algebra, letters are used to represent numbers. For example $a = 5$.

- To keep an **equation** balanced you must do the same to each side.

  $a + 4 = 7 \quad \rightarrow \quad a + 4 - 4 = 7 - 4 \quad \rightarrow \quad a = 3$

  $a - 3 = 1 \quad \rightarrow \quad a - 3 + 3 = 1 + 3 \quad \rightarrow \quad a = 4$

  $5a = 30 \quad \rightarrow \quad 5a \div 5 = 30 \div 5 \quad \rightarrow \quad a = 6$

  $\dfrac{a}{2} = 7 \quad \rightarrow \quad \dfrac{a}{2} \times 2 = 7 \times 2 \quad \rightarrow \quad a = 14$

- In a combined equation, deal with the $+$ and $-$ first.

  $3a + 7 = 1 \rightarrow 3a + 7 - 7 = 1 - 7 \rightarrow 3a = -6 \rightarrow a = -2$

- In an equation with **brackets**, **expand** the brackets first.

  $$3(x + 1) = 4 \quad \rightarrow \quad 3x + 3 = 4$$

- Quadratic equations can have 0, 1 or 2 solutions.

- You can find approximate solutions of more complicated equations by **trial and improvement**.

- $>$ means **greater than**

  $<$ means **less than**

  $\geqslant$ means **greater than or equal to**

  $\leqslant$ means **less than or equal to**

Shape, space and measure

2-D shapes

- The sum of the **exterior angles** of any polygon is 360°.

- The sum of the **interior angles** of a polygon with n sides is $(n \times 180°) - 360°$, or $(n - 2) \times 180°$

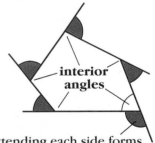

interior angles

extending each side forms the **exterior angles**

- A **bearing** is the angle measured from facing North and turning clockwise.

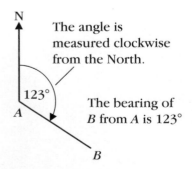

N

The angle is measured clockwise from the North.

123°

A

The bearing of B from A is 123°

B

2-D and 3-D shapes

- The **plan** of a solid is the view when seen from above.

- The **front elevation** is the view when seen from the front.

- The **side elevation** is the view when seen from the side.

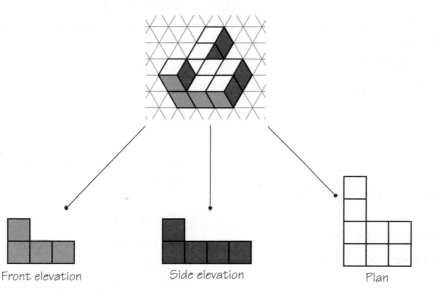

Front elevation   Side elevation   Plan

- Area of a circle $= A = \pi r^2$

Transformations

- You should be able to:

    – perform a rotation with a given centre and angle of rotation.

    – perform a reflection in a given mirror line.

    – describe rotations and reflections fully.

- You should be able to:

    – perform an enlargement with a given centre and scale factor.

    – perform a translation described by a column vector.

    – describe translations and enlargements fully.

Unit 3 Examination practice paper

A formula sheet can be found on page 161.

Section A (calculator)

1 The cost of having a car serviced is £56.40 before VAT at $17\frac{1}{2}$% is added.

Find the total cost after VAT is added.

(3 marks)

2 Work out the value of y.

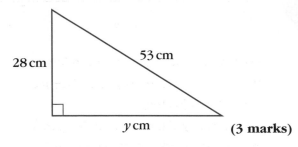

(3 marks)

3 The diameter of a wheel is 70 centimetres.

Work out how many revolutions the wheel makes when travelling 1 kilometre. **(4 marks)**

4 Here is a list of ingredients for making Lemon Surprise for 4 people:

2 lemons	2 eggs
50 g butter	250 ml milk
100 g sugar	50 g self raising flour

Work out how much of each ingredient is needed to make Lemon Surprise for 10 people. **(3 marks)**

5 (a) Find the value of n in $\dfrac{a^n \times a^3}{a^7} = a^2$

(b) Simplify $(2x^2y)^3$ **(4 marks)**

6 Reflect the shape P using the dotted line as the mirror line.

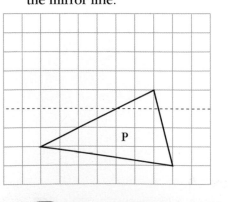

(2 marks)

7 $s = \frac{1}{2}(u + v)t$ is a formula linking distance, speed and time.

(a) Work out the value of s when $u = 0$, $v = 6$ and $t = 3$

(b) Make v the subject of the formula.

(5 marks)

8 Work out the area of a circle with radius 1.7 metres.

Give your answer correct to 3 significant figures. **(2 marks)**

9 $x^3 + 3x - 16 = 0$

Use trial and improvement to find the positive solution of this equation.

Give your answer correct to 1 decimal place.

(4 marks)

10 Convert the fraction $\frac{5}{12}$ to a decimal.

Give your answer correct to 2 decimal places.

(2 marks)

11 The total surface area, including the base, of this triangular prism is 2100 cm²

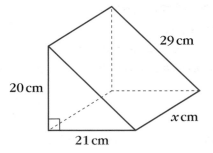

Work out the value of x. **(4 marks)**

12 Work out the value of $3y^2 - 2y$ when $y = -4$

(2 marks)

13 Construct a triangle ABC with $AB = 8$ cm, angle $A = 75°$ and angle $B = 30°$ **(2 marks)**

14 The length of a pencil is 120 mm.

Each week the pencil shortens by 5 mm.

Write down a formula for the length d mm, of the pencil after w weeks. **(3 marks)**

15 ABC is an equilateral triangle and $BCED$ is a square.

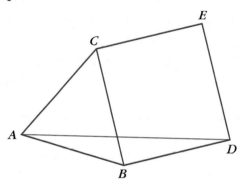

Work out the size of angle BDA.

Give reasons for your answer. **(4 marks)**

16 A ship sails so that it is equidistant from A and B until it is 3 km from C.

It then sails directly to C.

$AB = 6$ km, $BC = 10$ km, angle $ABC = 90°$

Draw a diagram with scale 1 cm = 1 km and construct the locus of the ship's path.

(5 marks)

17 This equilateral triangle is surrounded by regular polygons.

Work out the number of sides each polygon has.

(3 marks)

18 The graph shows part of a cyclist's trip to the seaside and back.

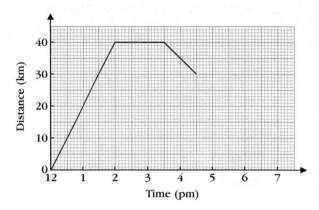

(a) How far was her total return journey?

(b) What was her average speed for the outward journey?

(c) How long did she stay at the seaside?

She completes her journey home at the same average speed as for her outward journey.

(d) What time does she get home?

(5 marks)

TOTAL FOR PAPER: 60 MARKS

END

Check your answers on page 173. For full worked solutions see the CD.

Section B (non-calculator)

1 Work out

    (a) $4 - (-6)$ **(b)** $20 - 4 + 5$ **(c)** 5×-3 **(d)** $12 \div (-0.2)$ **(4 marks)**

2 The cost of 15 paving stones is £36.

    Find the cost of 35 paving stones.   **(3 marks)**

3 (a) Write 65% as a fraction.

    (b) Write 6% as a decimal.

    (c) Write $\frac{3}{5}$ as a decimal.   **(3 marks)**

4 A packet of readimix cement contains sand and cement in the ration $4 : 1$.

    Find the amount of sand in a 10 kilogram bag of readimix.   **(4 marks)**

5 An electricity meter reading is 82375 units.

    The previous reading was 81265 units.

    The price per unit for electricity is 9.7 pence.

    Work out the cost of the electricity that has been used.   **(4 marks)**

6 (a) Solve $3x + 1 = 16$

    (b) Solve $\frac{2y}{5} = 3$

    (c) Solve $3(4 - 3x) = x + 7$ **(7 marks)**

7 Simplify

    (a) $a^5 \times a^2$ **(b)** $x^3 \div x^5$ **(c)** $15x^2y^4 \div 3xy^2$ **(4 marks)**

8 $-3 < x \leqslant 1$, where x is an integer.

    List all the possible values of x.   **(2 marks)**

9 (a) Find the mid-point of the line segment joining $(-1, -2)$ to $(1, 4)$.   **(4 marks)**

    (b) Find the gradient of the line that passes through these points.   **(4 marks)**

10 The perimeter of triangle ABC is 57 cm.

    Find the value of x.

    (4 marks)

11 Find $\frac{3}{5}$ of 215.   **(2 marks)**

12 A sports club has 240 junior members and 1240 senior members.

Express the number of junior members as a fraction of the total membership.

Give your answer in its simplest form. **(3 marks)**

13 (a) Simplify $\frac{3}{4} + \frac{4}{5}$ **(b)** Simplify $1\frac{3}{5} \times 2\frac{1}{4}$

Leave your answers as fractions in their simplest form. **(5 marks)**

14 Change $3\,\text{m}^3$ into cm^3. **(2 marks)**

15 Describe fully the transformation that maps P onto Q.

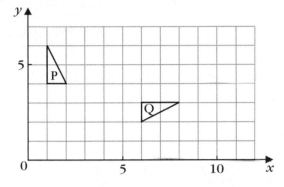

                                                                (4 marks)

16 Sketch the shape shown by the plan and elevations shown.

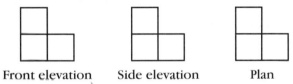

        Front elevation   Side elevation   Plan   **(2 marks)**

17 Given $57.1 \times 0.048 = 2.7408$

Find **(a)** 0.571×48 **(b)** $0.274\,08 \div 4.8$ **(2 marks)**

18 Draw the graph of $y = 2x^2 - 9x + 9$ for values of x from 0 to 5.

Use your graph to find estimates of the solution of $2x^2 - 9x + 9 = 2$ **(4 marks)**

TOTAL FOR PAPER: 60 MARKS

END

Check your answers on pages 173–174. For full worked solutions see the CD.

Formulae

Area of trapezium $= \frac{1}{2}(a + b)\,h$

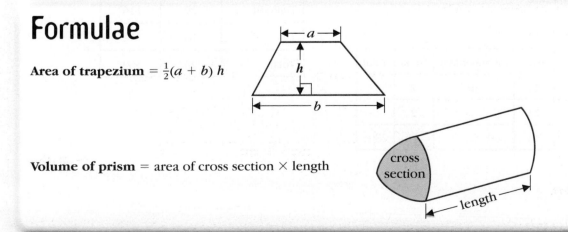

Volume of prism $=$ area of cross section $\times$ length

Answers

Unit 1

Collecting data

1 (a)

Drink	Tally	Frequency
Orange	IIII II	7
Grapefruit	III	3
Cranberry	IIII	4
Tropical	IIII I	6

(b) 3

(c) Orange

2 (a) Continuous

(b) Discrete

(c) Discrete

(d) Continuous

3 (a) Options are too vague; it is not possible to answer 'none'; no time period specified.

(b) How much money do you spend in the café in a week?

£0 ☐   £0.01−£4.99 ☐   £5−£9.99 ☐

£10−£20 ☐   more than £20 ☐

4

Type of pet	Tally	Frequency
Dog		
Cat		
Hamster		
Rabbit		
Goldfish		

Organising data

1 (a) Mary

(b) Jackie, Sakina

(c) Jackie

2

	Comedy	Soap	Documentary	News	Total
Men	7	11	10	9	37
Women	12	23	6	2	43
Total	19	34	16	11	80

Charts

1 (a) (i) 30

(ii) 25

(b)

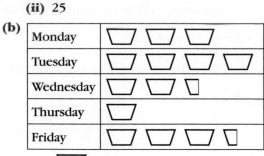

Key: ⬓ represents 10 teas

2 (a) 25

(b) Science

(c) History

(d) 80

3

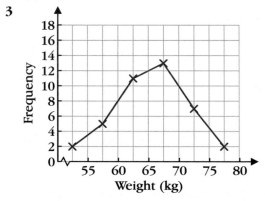

Pie charts and stem and leaf diagrams

1

Flower	Number	Angle
Snowdrop	23	92°
Crocus	20	80°
Daffodil	29	116°
Lily	18	72°
Total	90	360°

2

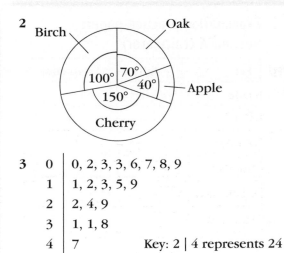

Birch, Oak, Apple, Cherry pie chart with 100°, 70°, 40°, 150°

3

0	0, 2, 3, 3, 6, 7, 8, 9
1	1, 2, 3, 5, 9
2	2, 4, 9
3	1, 1, 8
4	7

Key: 2 | 4 represents 24

Time series and scatter graphs

1 (a)

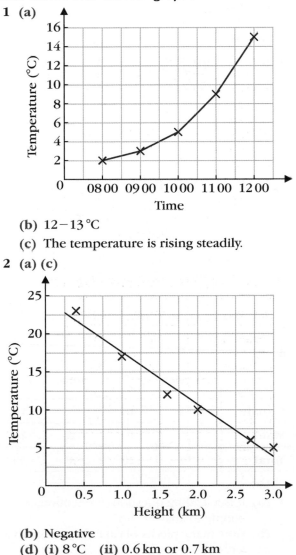

(b) 12−13 °C

(c) The temperature is rising steadily.

2 (a) (c)

(b) Negative

(d) (i) 8 °C **(ii)** 0.6 km or 0.7 km

Averages and the range

1 (a) 33 **(b)** 30 **(c)** 10 **(d)** 31

2 (a) 34 **(b)** 2 **(c)** 2 **(d)** 2.12

(e) The most tries the team scored in any match was 4, so the average cannot be more than 4.

3 (a) $20 \leqslant t < 25$

(b) $20 \leqslant t < 25$

(c) 23.5 minutes

Probability

1

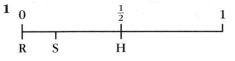

2 0.35

3 (a) (B1, R1), (B1, R2), (B1, R3), (B1, R4),
(B1, R5), (B1, R6)
(B2, R1), (B2, R2), (B2, R3), (B2, R4),
(B2, R5), (B2, R6)
(B3, R1), (B3, R2), (B3, R3), (B3, R4),
(B3, R5), (B3, R6)
(B4, R1), (B4, R2), (B4, R3), (B4, R4),
(B4, R5), (B4, R6)
(B5, R1), (B5, R2), (B5, R3), (B5, R4),
(B5, R5), (B5, R6)
(B6, R1), (B6, R2), (B6, R3), (B6, R4),
(B6, R5), (B6, R6)

(b) $\frac{6}{36} = \frac{1}{6}$

4 $\frac{3}{6} = \frac{1}{2}$

5 (a)

	Car	Walk	Bus	Train	Total
Men	12	2	3	7	24
Women	16	7	2	1	26
Total	28	9	5	8	50

(b) (i) $\frac{8}{50}$ **(ii)** $\frac{2}{50}$ **(iii)** $\frac{16}{50}$

Unit 1 Handling data: subject test

1 (a)

Sport	Tally	Frequency
Soccer	⊬⊦⊦ ||||	9
Golf	|||	3
Rugby	⊬⊦⊦		6
Cricket	||	2

(b) 6 **(c)** Soccer

2 (a) (i) 80   **(ii)** 50

(b)

Saturday	⊕ ⊕ ⊕ ⊕
Sunday	⊕ ⊕ ◖
Monday	⊕ ⊕
Tuesday	⊕ ◖

Key: ⊕ represents 10 teas

3 (a) 8   **(b)** 7   **(c)** 4

4

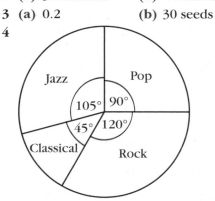

5 (a)

```
2 | 7, 8
3 | 3, 6, 7, 9, 9
4 | 1, 4, 5, 6, 6, 8, 9
5 | 0, 1, 3
6 | 0, 2, 5     Key: 2 | 7 means 27 seconds
```

(b) 45.5 seconds

6 (a)

	Car	Walk	Cycle	Total
Year 7	19	13	9	41
Year 8	5	12	5	22
Year 9	12	18	7	37
Total	36	43	21	100

(b) (i) $\frac{9}{25}$ **(ii)** $\frac{9}{100}$

7 0.25

8 (a) (c)

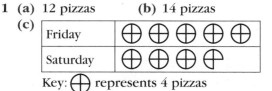

Weight (kg)

(b) Positive   **(d) (i)** 77 kg   **(ii)** 184 cm

Unit 1 Examination practice paper: Section A (calculator)

1 (a)

Pet	Tally	Frequency
Dog	⊮ III	8
Cat	IIII	4
Rabbit	II	2
Hamster	II	2
Fish	IIII	4

(b) Dog

(c) Cat and fish

2 (a) 30 matches   **(b)** 4 matches

3 (a) 0.2   **(b)** 30 seeds

4

Unit 1 Examination practice paper Section B (non-calculator)

1 (a) 12 pizzas   **(b)** 14 pizzas

(c)

Friday	⊕ ⊕ ⊕ ⊕ ⊕
Saturday	⊕ ⊕ ⊕ ◖

Key: ⊕ represents 4 pizzas

2 (a) 100 ice-creams   **(b)** 150 ice-creams

(c) The weather might have been wet and cold.

3 (a)

	Sandwich	School lunch	Home	Total
Female	5	10	1	16
Male	2	10	2	14
Total	7	20	3	30

(b) $\frac{1}{3}$

4 (a) It is a leading question; it encourages people to answer 'Yes'.

(b) How many pieces of cake do you eat in a week?

0 ☐   1–2 ☐   3–5 ☐   6–8 ☐

more than 8 ☐

Unit 2

Place value, ordering and rounding

1 Seventy-five thousand, two hundred and three
2 17 354 000
3 800
4 39, 72, 88, 267, 302
5 (a) 57 000   (b) 56 800

Negative numbers

1 (a) 10   (b) −10, −6, −2, 0, 2, 6, 10
2 (a) −1   (b) −11   (c) 1
   (d) −1   (e) 11   (f) 1
3 2 °C
4 (a) −20   (b) −20   (c) 20   (d) −4
   (e) −4   (f) 4   (g) 4   (h) −5

Indices and powers

1 (a) 27   (b) 6, −6
2 (a) 500   (b) 15
3 $3 \times (4 + 5) = 3^3$
4 (a) 4^5 (b) 9^3 (c) 1   (d) 4^6

Multiples, factors and primes

1 (a) 4, 8 and 16   (b) 4, 8 and 16
   (c) 3 and 5
2 (a) $16 = 2 \times 2 \times 2 \times 2$
   (b) $24 = 2 \times 2 \times 2 \times 3$
3 8
4 48

Calculating and estimating

1 (a) 6970   (b) 28 938   (c) 13 300
2 (a) 25   (b) 24 r8   (c) 31
3 (a) 3   (b) 50
4 (a) 2.973…   (b) 55.233…

Adding, subtracting, multiplying and dividing decimals

1 (a) 15.65   (b) 16.05   (c) £22.85
2 (a) 0.76   (b) 9.15   (c) £2.76
3 (a) 37.8   (b) 66.15   (c) 2.1452
4 (a) 9   (b) 3.2   (c) 12

Rounding decimals

1 Rana £10.50, Axel £8.06
2 £0.90 or 90p
3 (a) (i) 5.5   (ii) 10.4   (iii) 4.1
   (b) (i) 5.45   (ii) 10.40   (iii) 4.06

4 (a) (i) 300 000   (ii) 60 000
       (iii) 0.3   (iv) 0.0006
   (b) (i) 250 000   (ii) 56 900
       (iii) 0.347   (iv) 0.000 600

Fractions and percentages

1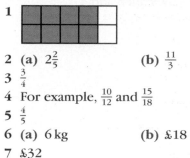
2 (a) $2\frac{2}{5}$ (b) $\frac{11}{3}$
3 $\frac{3}{4}$
4 For example, $\frac{10}{12}$ and $\frac{15}{18}$
5 $\frac{4}{5}$
6 (a) 6 kg   (b) £18
7 £32

Unit 2 Number: subject test

1 (a) Seven thousand four hundred and thirty two
   (b) 23 000
   (c) 3 thousands or 3000
   (d) 24 576
2 $\frac{3}{4}$
3 (a) 17   36   42   84   101
   (b) 0.006   0.06   0.09   0.6   0.69
   (c) −7   −3   −2   0   3   7
4 (a) 8   (b) −10   (c) 6
5 (a) 11   (b) 20
6 (a) 10 578   (b) 25
7 (a) 72   (b) 12
8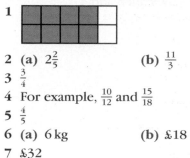
9 £28
10 (a) 64   (b) 1.3
11 (a) 7500   (b) 1.2876…
    (c) (i) 1.29   (ii) 1
12 (a) 7^8 or 5 764 801
    (b) 5^4 or 625

Simplifying algebra

1 (a) $5e$ (b) $3jk$
2 (a) $8a + b$ (b) $4p^3$
3 (a) $15pq$ (b) t^3
4 (a) $2c + 12$ (b) $ad − 3d$ (c) $b^2 + 2b$
5 (a) $22p + 2$ (b) $3g + 13h$ (c) $7p + 6$
6 (a) $3(c + 4)$ (b) $4m(m − 3)$ (c) $2t(t − 3)$

Number sequences

1 (a)

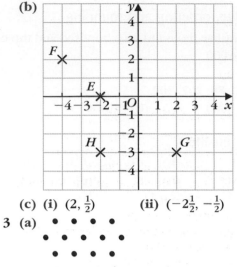

Diagram 4

(b) 16, 19
(c) Multiply 15 by 3, then add 4
(d) $3n + 4$

2 22, 18

3 $7n - 3$

Coordinates

1 (a) $(3, 2)$
   (b) $(0, 3)$

2 (a) $(-2, -3)$
   (b) $(-3, 1)$

3 $(4, 2\frac{1}{2})$

4 $K(3, 3, 0), L(3, 1, 0), M(3, 1, 2), N(0, 1, 2)$

Algebraic line graphs

1

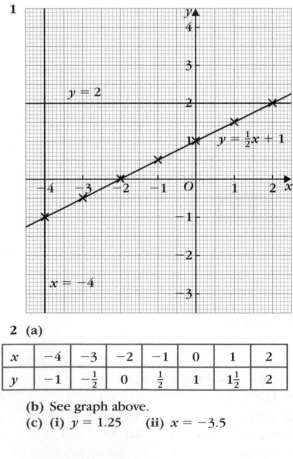

2 (a)

x	-4	-3	-2	-1	0	1	2
y	-1	$-\frac{1}{2}$	0	$\frac{1}{2}$	1	$1\frac{1}{2}$	2

(b) See graph above.
(c) (i) $y = 1.25$ **(ii)** $x = -3.5$

Unit 2 Algebra: subject test

1 17, 21

2 (a) $A(3, -1); B(1, 2); C(-3, -4); D(-2, 3)$

(b)

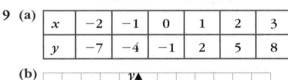

(c) (i) $(2, \frac{1}{2})$ **(ii)** $(-2\frac{1}{2}, -\frac{1}{2})$

3 (a)

Diagram 4

(b)

Diagram	1	2	3	4	5
Number of dots	4	7	10	13	16

(c) $3n + 1$

4 $3n + 5$

5 $A(0, 2, 0); B(3, 0, 2); C(0, 0, 2); D(3, 2, 2)$

6 $2k$ **7** $4ab$ **8** $3e^2$

9 (a)

x	-2	-1	0	1	2	3
y	-7	-4	-1	2	5	8

(b)

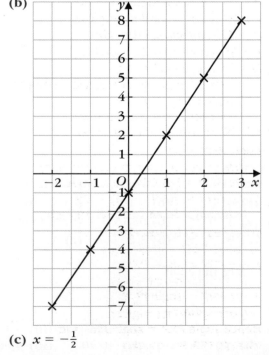

(c) $x = -\frac{1}{2}$

10 $8pq$

11 g^4

12 $3a - 6b$

13 $6c - 2d$

14 $3(w - 4)$

15 $x(x - 5)$

Naming and calculating angles

1 $x = 113°$ (angles at a point)

2 $w = 25°$ (vertically opposite angles)
   $x = 85°$ (angles on a straight line)
   $y = 155°$ (angles on a straight line, or angles at a point)

3 (a) 150° Angle $DCB = 30°$
       (base angle in an isosceles triangle)
       Angle $DCF = 150°$
       (angles on a straight line)
   (b) 48° Angle $DBC = 30°$
       (base angle in an isosceles triangle)
       Angle $DBA = 150°$
       (angles on a straight line)
       Angle $EAB = 48°$
       (angles in a quadrilateral)
   (c) 90° Angle $DCB = 30°$
       (base angle in an isosceles triangle)
       Angle $BCG = 60°$
       (angle in an equilateral triangle)

Working with angles

1 $a = 47°$ (alternate angles),
   $b = 47°$ (corresponding angles),
   $c = 84°$ (corresponding angles),
   $d = 37°$ (angles in a triangle)

2 (a) 35° (corresponding angles)
   (b) 70° Angle $ABE = 110°$
       (third angle in an isosceles triangle)
       Angle $CBE = 70°$
       (angles on a straight line)
   (c) 70° (corresponding angles)

3 Angle $ABC = 34°$
   (base angle in an isosceles triangle)
   Angle $BCD = 34°$ (alternate angles)
   Angle $CDB = \frac{1}{2}(180 - 34) = 73°$
   (base angle in an isosceles triangle)

4 Triangle PXQ is isosceles.
   So angle PXQ = angle PQX
   Angle PYR = angle PXQ
     (corresponding angles)
   Angle PRY = angle PQX
     (corresponding angles)
   Hence angle PYR = angle PRY and
   triangle PRY is isosceles.

Units of measurement

1 (a) 09:15 (b) 34 minutes (c) 06:44

2 (a) 7:24 am (b) 275 g

3 (a) 17.6 pounds (b) 104 km
   (c) 30 inches

4 12 km/h

5 (a) 126 km (b) 56 km (c) 7 km

6 1000 metres

Perimeter and area

1 Perimeter 20 cm, area 18 cm²

2 40 cm²

3 32 cm²

4 44 cm²

5 550 cm²

6 570 cm²

Volume, capacity and density

1 270 litres

2 480 cm³

3 192 cm³

4 7 cm

5 (a) 160 cm³ (b) 416 g

6 (a) 19 g/cm³ (b) 19 000 kg/m³

7 (a) 1884.96 cm³ (b) 3078.76 cm³

Unit 2 Shape, space and measure: subject test

1 (a) Perimeter 16 cm, area 8 cm²
   (b) Perimeter 56 cm, area 170 cm²

2 25°

3 (a) 53 cm² (b) 26 cm² (c) 18 cm²

4 45°

5 (a) $a = 46°$ (b) $c = 105°$
   (c) $x = 43°, y = 70°$

6 20 cubes

7 50 km/h

8 (a) 50 s
   (b) 0.1 hours *or* 6 minutes

9 1180 cm³

10 1400

11 (a) 80° (b) 53° (c) 47° (d) 53°

12 520 cm²

13 45 km

14 768 g

Unit 2 Examination practice paper
Stage 1

1 E	**2** D	**3** D	**4** B
5 D	**6** B	**7** B	**8** B
9 C	**10** D	**11** A	**12** A
13 D	**14** D	**15** C	**16** E
17 D	**18** E	**19** B	**20** B
21 B	**22** D	**23** D	**24** A
25 E			

Unit 2 Examination practice paper
Stage 2

1 (a) $27\,cm^3$ **(b)** $80\,cm^3$

2 30 people

3 (a) 121   **(b)** 1.3
  (c) -3 **(d)** -3

4 (a) 27 cm   **(b)** Size 11   **(c)** 26.5 cm

5 (a) 80   **(b)** 45

6 (a) $6b + 15$ **(b)** $5(g + 3)$
  (c) $x^2 + 7x + 10$

7 12.5 cm

8 (a) $3^6 = 729$ **(b)** 4

9 180 boxes

Unit 3

Working with fractions

1 £24

2 (a) $1\frac{1}{12}$ **(b)** $8\frac{5}{24}$

3 (a) $\frac{13}{24}$ **(b)** $3\frac{5}{12}$

4 (a) $\frac{7}{16}$ **(b)** 5

5 (a) $1\frac{7}{18}$ **(b)** $\frac{2}{3}$

6 $1\frac{7}{12}$ miles

Percentages, fractions and decimals

1 $\frac{3}{5}$, 63%, 0.65, $\frac{2}{3}$, 67%

2 (a) 9 kg   **(b)** £24

3 £48

4 £15

5 She did equally well (75%) in French and Spanish but worse (67%) in German.

Using percentages

1 (a) £76.50   **(b)** £63.75

2 £188

3 (a) £20   **(b)** £60

4 20%

Ratio

1 (a) $3:2$ **(b)** $2:1$ **(c)** $3:2$ **(d)** $4:3$

2 $6:4$ and $12:8$

3 (a) (i) $1.25:1$ **(ii)** $0.667:1$
  (b) (i) $1:0.8$ **(ii)** $1:1.5$

4 (a) $\frac{5}{8}$ **(b)** Wayne £15, Tracey £9

5 £1000

Proportion

1 £4.00   **2 (a)** 5 eggs   **(b)** 300 g

3 1 km

Unit 3 Number: subject test

1 44 bars

2 65%, $\frac{2}{3}$, $\frac{4}{5}$, 0.81, 82%

3 £60

4 (a) £50   **(b)** £52   **(c)** £58.75

5 (a) £14   **(b)** £75.20

6 £56

7 £259.65

8 4.8 hours *or* 4 hours 48 minutes

9 (a) 18 bags   **(b)** 8 bags

10 3350 m *or* 3.35 km

11 (a) £27   **(b)** $\frac{1}{2}$, $\frac{3}{5}$, $\frac{2}{3}$, $\frac{3}{4}$
  (c) $\frac{5}{12}$ **(d)** 8

12 Marco £20, José £12

Algebra

1 t^3

2 (a) h^7 **(b)** $4x^3$

3 (a) $4x^2$ **(b)** $2y^2$

4 (a) $3q^4$ **(b)** $4t^3$

5 (a) a^{12} **(b)** $3y$ **(c)** n^4

6 $8m^3n^2$

Linear graphs

1 (a) 1 hour   **(b)** 45 km   **(c)** 60 km/h

2 (a) 40 m   **(b)** 10 seconds   **(c)** 3 m/s

3 (a) 60 km/h
  (b)

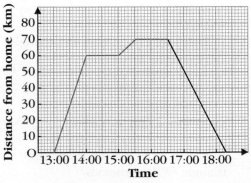

Curved graphs

1 (a)

x	-1	0	1	2	3	4
y	4	0	-2	-2	0	4

(b)

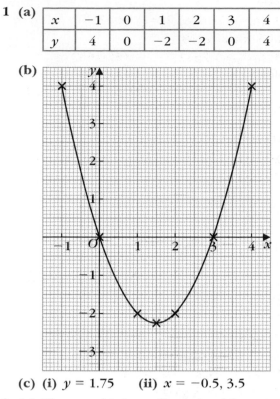

(c) (i) $y = 1.75$ **(ii)** $x = -0.5, 3.5$

2 (a) The sound is immediately loud for a time.
  (b) The sound decreases at a constant rate.
  (c) The sound level remains constant.

Formulae

1 180
2 £11
3 $W = 8n + kn$
4 $P = 13y + 3$

Rearranging formulae

1 (a) 4   **(b)** -0.4

2 (a) $x = \dfrac{y + 4}{3}$ **(b)** $x = \dfrac{y - 3}{2}$

3 (a) $x = \dfrac{a - bc}{4}$ **(b)** $x = \dfrac{t - 6}{2}$

4 2.5

5 $x = 2(y - 8)$

Linear equations

1 -9 **2** $6\frac{1}{2}$ **3** -1
4 $4\frac{1}{2}$ **5** $3\frac{1}{2}$ **6** $15\frac{1}{3}$

Solving non-linear equations and inequalities

1 ± 7
2 2.4

3

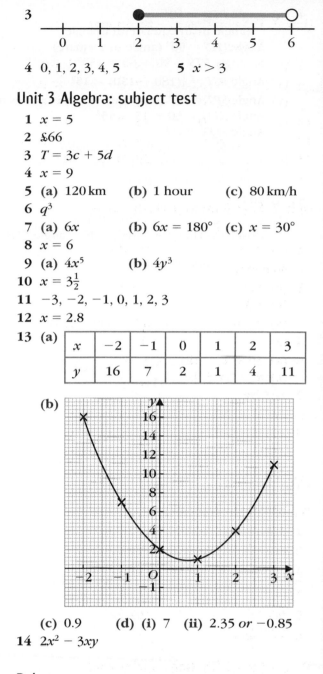

4 0, 1, 2, 3, 4, 5   **5** $x > 3$

Unit 3 Algebra: subject test

1 $x = 5$
2 £66
3 $T = 3c + 5d$
4 $x = 9$
5 (a) 120 km   **(b)** 1 hour   **(c)** 80 km/h
6 q^3
7 (a) $6x$ **(b)** $6x = 180°$ **(c)** $x = 30°$
8 $x = 6$
9 (a) $4x^5$ **(b)** $4y^3$
10 $x = 3\frac{1}{2}$
11 $-3, -2, -1, 0, 1, 2, 3$
12 $x = 2.8$

13 (a)

x	-2	-1	0	1	2	3
y	16	7	2	1	4	11

(b)

(c) 0.9   **(d) (i)** 7   **(ii)** 2.35 *or* -0.85

14 $2x^2 - 3xy$

Polygons

1 Triangles A and C
2 (a)

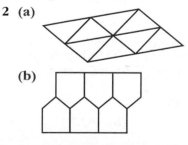

(b)

3 $b = 75°$ (angles on a straight line),
   $a = 65°$ (exterior angles sum to 360°)

4 (a) Angle $PTQ = 60°$
   (angle in an equilateral triangle)
   Angle $QTS = 90°$ (angle in a square)
   So angle $PTS = 90 + 60 = 150°$

(b) Angle $PST = \frac{1}{2}(180 - 150) = 15°$

(c) Angle $SPT = 15°$ (base angle in an isosceles triangle)
   Angle $XPQ = 60 - 15 = 45°$
   Angle $PQX = 60°$
     (angle in an equilateral triangle)
   Angle $PXQ = 180 - 60 - 45 = 75°$
     (angles in a triangle)

5 (a) $105°$

(b) $CD = BC = CE$ so triangle BCE is
   isosceles. Therefore angle CBE = angle CEB

Drawing and calculating

1

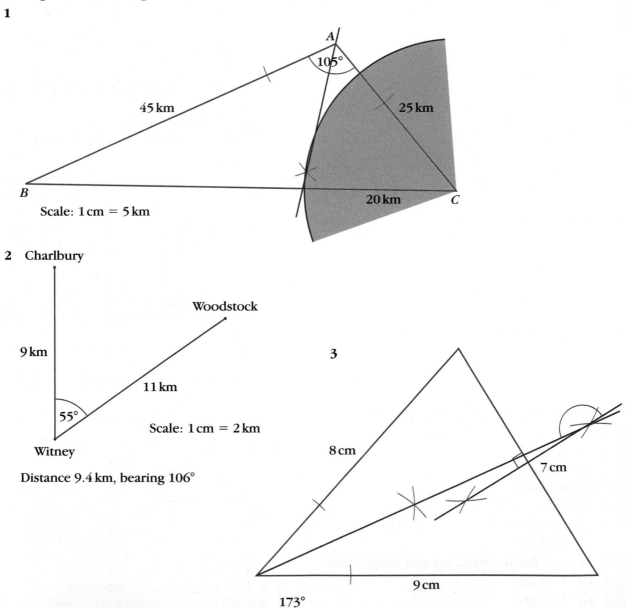

Scale: 1 cm = 5 km

2 Charlbury

Distance 9.4 km, bearing 106°

Scale: 1 cm = 2 km

3

4 (a) 53 cm
  (b) 60 cm
  (c) 12.6 cm

3-D shapes

1 (a)

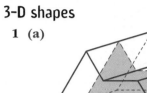

(b)

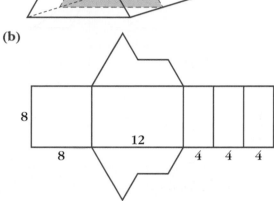

(c) Plan   Side   Front
                 elevation   elevation

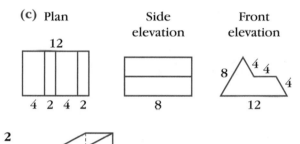

2

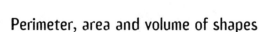

Perimeter, area and volume of shapes

1 Area $1.13\,\text{m}^2$, circumference $3.77\,\text{m}$
2 58 cm
3 (a) $3619\,\text{cm}^3$
  (b) $2413\,\text{cm}^3$
4 1.81 cm
5 (a) $1508\,\text{cm}^2$
  (b) $1005\,\text{cm}^2$
6 Radius 2.65 cm, area of cross-section $22.1\,\text{cm}^2$
7 (a) $1257\,\text{mm}^3$
  (b) $7.96\,\text{g/cm}^3$

Rotation, reflection and symmetry

1

2 (a)

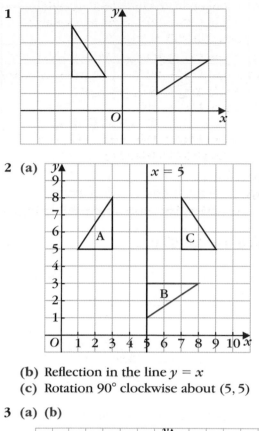

  (b) Reflection in the line $y = x$
  (c) Rotation 90° clockwise about $(5, 5)$

3 (a) (b)

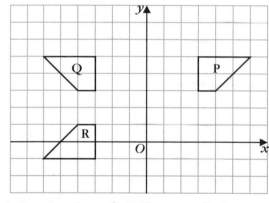

  (c) Rotation through 180°, centre $(0, 2)$

Translations and enlargements

1

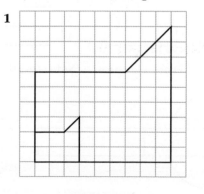

2

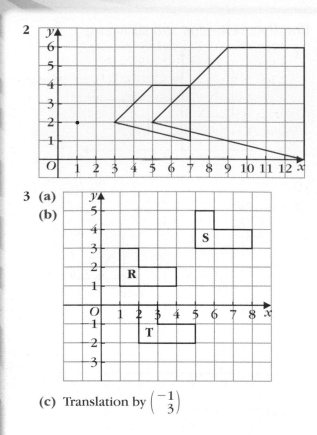

3 (a)
(b)

(c) Translation by $\begin{pmatrix} -1 \\ 3 \end{pmatrix}$

Unit 3 Shape, space and measure: subject test

1 A and F, B and E, C and D, G and H

2 (a) $x = 110°$ **(b)** $a = 65°$

3

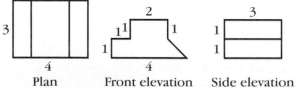

4

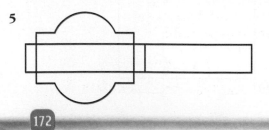

Plan   Front elevation   Side elevation

5

6 78 540 cm *or* 785.4 m

7 (a)

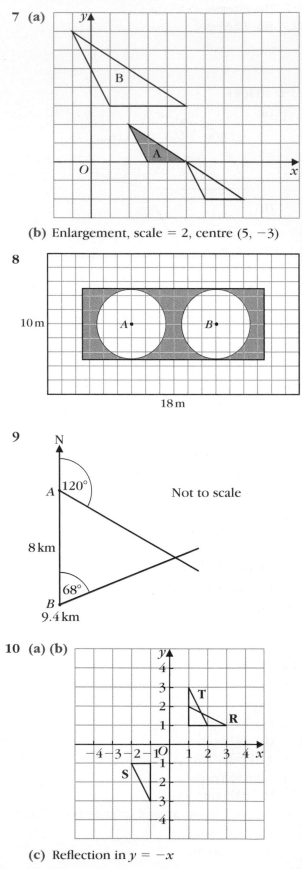

(b) Enlargement, scale = 2, centre (5, −3)

8

9

10 (a) (b)

(c) Reflection in $y = -x$

11 $BXC = 36°$

12 Diameter = 15.96 cm

13 $AC = 20.4$ cm

Unit 3 Examination practice paper
Section A (calculator)

1 £66.27

2 $y = 45$

3 454.7 revolutions (to 1 d.p.)

4 Lemon surprise (for 10)

5 lemons

125 g butter

250 g sugar

5 eggs

625 ml milk

125 g self raising flour

5 (a) $n = 6$

  (b) $8x^6y^3$

6

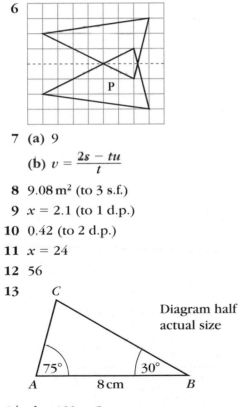

7 (a) 9

  (b) $v = \dfrac{2s - tu}{t}$

8 9.08 m² (to 3 s.f.)

9 $x = 2.1$ (to 1 d.p.)

10 0.42 (to 2 d.p.)

11 $x = 24$

12 56

13

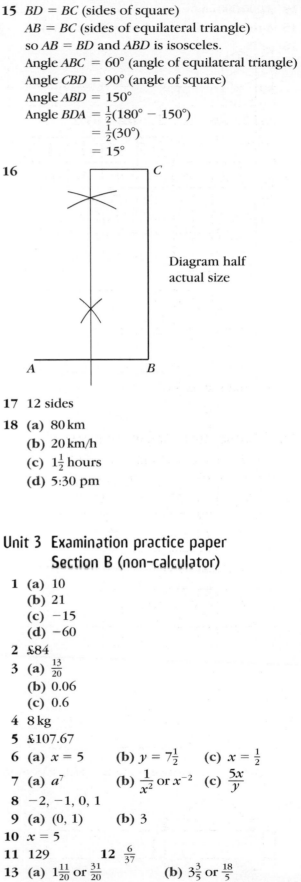

Diagram half actual size

C, 75°, A, 8 cm, 30°, B

14 $d = 120 - 5w$

15 $BD = BC$ (sides of square)

  $AB = BC$ (sides of equilateral triangle)

  so $AB = BD$ and ABD is isosceles.

  Angle $ABC = 60°$ (angle of equilateral triangle)

  Angle $CBD = 90°$ (angle of square)

  Angle $ABD = 150°$

  Angle $BDA = \frac{1}{2}(180° - 150°)$

    $= \frac{1}{2}(30°)$

    $= 15°$

16

Diagram half actual size

17 12 sides

18 (a) 80 km

  (b) 20 km/h

  (c) $1\frac{1}{2}$ hours

  (d) 5:30 pm

Unit 3 Examination practice paper
Section B (non-calculator)

1 (a) 10

  (b) 21

  (c) -15

  (d) -60

2 £84

3 (a) $\frac{13}{20}$

  (b) 0.06

  (c) 0.6

4 8 kg

5 £107.67

6 (a) $x = 5$ (b) $y = 7\frac{1}{2}$ (c) $x = \frac{1}{2}$

7 (a) a^7 (b) $\dfrac{1}{x^2}$ or x^{-2} (c) $\dfrac{5x}{y}$

8 $-2, -1, 0, 1$

9 (a) $(0, 1)$ (b) 3

10 $x = 5$

11 129   **12** $\frac{6}{37}$

13 (a) $1\frac{11}{20}$ or $\frac{31}{20}$ (b) $3\frac{3}{5}$ or $\frac{18}{5}$

14 3 000 000 cm³

15 A rotation of 90° clockwise about (3, 1)

16

17 (a) 27.408   (b) 0.0571

18

x	0	1	2	3	4	5
y	9	2	−1	0	5	14

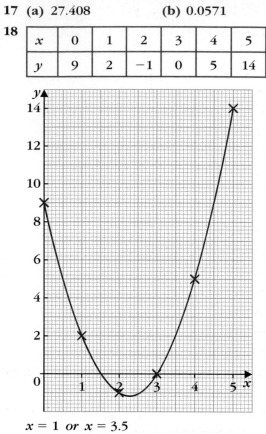

$x = 1$ *or* $x = 3.5$

Index